Dingyü Xue
**MATLAB**® Programming

## Also of Interest

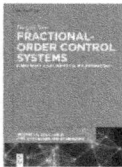

*Fractional-Order Control Systems, Fundamentals and Numerical Implementations*
Dingyü Xue, 2017
ISBN 978-3-11-049999-5, e-ISBN (PDF) 978-3-11-049797-7,
e-ISBN (EPUB) 978-3-11-049719-9

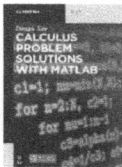

*Calculus Problem Solutions with MATLAB®*
Dingyü Xue, 2020
ISBN 978-3-11-066362-4, e-ISBN (PDF) 978-3-11-066697-7,
e-ISBN (EPUB) 978-3-11-066375-4

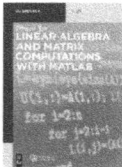

*Linear Algebra and Matrix Computations with MATLAB®*
Dingyü Xue, 2020
ISBN 978-3-11-066363-1, e-ISBN (PDF) 978-3-11-066699-1,
e-ISBN (EPUB) 978-3-11-066371-6

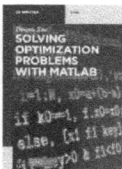

*Solving Optimization Problems with MATLAB®*
Dingyü Xue, 2020
ISBN 978-3-11-066364-8, e-ISBN (PDF) 978-3-11-066701-1,
e-ISBN (EPUB) 978-3-11-066369-3

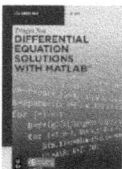

*Differential Equation Solutions with MATLAB®*
Dingyü Xue, 2020
ISBN 978-3-11-067524-5, e-ISBN (PDF) 978-3-11-067525-2,
e-ISBN (EPUB) 978-3-11-067531-3

Dingyü Xue

# MATLAB® Programming

Mathematical Problem Solutions

DE GRUYTER

清华大学出版社
TSINGHUA UNIVERSITY PRESS

**Author**
Prof. Dingyü Xue
School of Information Science and Engineering
Northeastern University
Wenhua Road 3rd Street
110819 Shenyang
China
xuedingyu@mail.neu.edu.cn

ISBN 978-3-11-066356-3
e-ISBN (PDF) 978-3-11-066695-3
e-ISBN (EPUB) 978-3-11-066370-9

**Library of Congress Control Number: 2019955271**

**Bibliographic information published by the Deutsche Nationalbibliothek**
The Deutsche Nationalbibliothek lists this publication in the Deutsche Nationalbibliografie; detailed bibliographic data are available on the Internet at http://dnb.dnb.de.

© 2020 Tsinghua University Press Limited and Walter de Gruyter GmbH, Berlin/Boston
Cover image: Dingyü Xue
Typesetting: VTeX UAB, Lithuania
Printing and binding: CPI books GmbH, Leck

www.degruyter.com

# Preface

Scientific computing is commonly and inevitably encountered in course learning, scientific research and engineering practice for each scientific and engineering student and researcher. For the students and researchers in the disciplines which are not pure mathematics, it is usually not a wise thing to learn thoroughly low-level details of related mathematical problems, and also it is not a simple thing to find solutions of complicated problems by hand. It is an effective way to tackle scientific problems, with high efficiency and in accurate and creative manner, with the most advanced computer tools. This method is especially useful in satisfying the needs for those in the area of science and engineering.

The author had made some effort towards this goal by addressing directly the solution methods for various branches in mathematics in a single book. Such a book, entitled "MATLAB based solutions to advanced applied mathematics", was published first in 2004 by Tsinghua University Press. Several new editions are published afterwards: in 2015, the second edition in English by CRC Press, and in 2018, the fourth edition in Chinese were published. Based on the latest Chinese edition, a brand new MOOC project was released in 2018,[1] and received significant attention. The number of the registered students was about 14 000 in the first round of the MOOC course, and reached tens of thousands in later rounds. The textbook has been cited tens of thousands times by journal papers, books, and degree theses.

The author has over 30 years of extensive experience of using MATLAB in scientific research and education. Significant amount of materials and first-hand knowledge has been accumulated, which cannot be covered in a single book. A series entitled "Professor Xue Dingyü's Lecture Hall" of such works are scheduled with Tsinghua University Press, and the English editions are included in the DG STEM series with De Gruyter. These books are intended to provide systematic, extensive and deep explorations in scientific computing skills with the use of MATLAB and related tools. The author wants to express his sincere gratitude to his supervisor, Professor Derek Atherton of Sussex University, who first brought him into the paradise of MATLAB.

This MATLAB series is not a simple revision of the existing books. With decades of experience and material accumulation, the idea of "revisiting" is adopted in authoring these books, in contrast to other mathematics and other MATLAB-rich books. The viewpoint of an engineering professor is established and the focus is on solving various applied mathematical problems with tools. Many innovative skills and general-purpose solvers are provided to solve problems with MATLAB, which is not possible by any other existing solvers, so as to better illustrate the applications of computer tools in solving mathematical problems in every mathematics branch. It also helps

---

1 MOOC (in Chinese) address: https://www.icourse163.org/learn/NEU-1002660001

https://doi.org/10.1515/9783110666953-201

the readers broaden their viewpoints in scientific computing, and even in finding innovative solutions by themselves to scientific computing which cannot be solved by any other existing methods.

The first title in the MATLAB series can be used as an entry-level textbook or reference book to MATLAB programming, so as to establish a solid foundation and deep understanding for the application of MATLAB in scientific computing. Each subsequent volume tries to cover a branch or topic in mathematical courses. These MATLAB books are suitable for the readers who have already learnt the related mathematical courses, and revisit the courses to learn how to solve the problems by using computer tools. It can also be used as a companion in synchronizing the learning of related mathematics courses, and for viewing the course from a different angle. The readers may expand their knowledge in learning the related courses, so as to better understand and practice the relevant materials. Bearing in mind the "computational thinking" in authoring the series, deep understanding and explorations are made for each mathematics branch involved.

This book is the first in the MATLAB series. Systematic illustration of MATLAB programming is provided in the book. The commonly used data and statement structures in MATLAB are introduced first, followed by algebraic and transcendental function evaluations of matrices, and data manipulations. Flow controls and applications in MATLAB programming are then introduced, and MATLAB function programming and skills are provided. Scientific visualization in MATLAB are addressed. Advanced topics such as the designs of MATLAB interface to other languages, object-oriented programming and graphical user interface are illustrated, such that a better foundation can be established for the readers to continue learning scientific computing and the knowledge of other related mathematical fields.

At the time the books are published, the author wishes to express his sincere gratitude to his wife, Professor Yang Jun. Her love and selfless care over the decades provided the author immense power, which supports his academic research, teaching and writing.

September 2019                                                                                   Xue Dingyü

# Contents

# 1 Introduction to computer mathematics languages

## 1.1 Introduction to solving mathematical problems

Mathematical problems are essential in almost all aspects of scientific and engineering research. The mathematical models should normally be established first, then the solutions of the models under investigation can be obtained. Specific knowledge is required for the establishment of mathematical models, which needs the expertise of the researchers, and with the established models, the numerical and analytical approaches presented in this series can be used to solve the problems. In this chapter, a brief introduction to computer mathematical problems is proposed. It will be illustrated through simple examples why computer mathematics languages should be learned. A concise history of the development of computer mathematics languages and mathematical tools will be introduced.

### 1.1.1 Why learn a computer mathematics language?

It is well known that manual formulation of mathematical problems is a useful skill if the problems are not too complicated. Unfortunately, for a great variety of mathematical problems in real world, manual formulation is laborious or even impossible. Computer tools must be employed for tackling these problems.

There are essentially two ways for solving mathematical problems with computers. One is to implement the existing numerical algorithms with general-purpose computer languages, such as C and Fortran. The other way is to use specific computer languages with good reputation. Appropriate computer languages such as MATLAB, Mathematica[17], and Maple[13] can be adopted. In this book, the languages are referred to as the computer mathematics languages. Numerical algorithms can only be used to tackle computation problems with numbers, while for problems like finding the solutions to the quadratic equation $ax^2 + bx + c = 0$, where $a$, $c$, $d$ are not given numerical values but symbolic variables, numerical algorithms, no matter how sophisticated, cannot be adopted. The professional computer mathematics languages capabilities should be adopted instead.

The term "mathematical computation" will be used throughout the series, whereas the term really means both numerical and analytical computation of solutions to mathematical problems. Normally, analytical solutions are explored first, and if there are no analytical solutions, numerical solutions are obtained.

The following examples are shown so that the readers may understand and appreciate the necessity of using the computer mathematics languages.

https://doi.org/10.1515/9783110666953-001

**Example 1.1.** Consider a simple problem. What is the last digit of $2019^{2019}$?

**Solutions.** If computers are not used, all the mathematicians know about the result is its last digit, since tricks can be used in finding it. In fact, the solution is useless in reality, since if you spend some money to buy something, you will never care what the last digit is. It makes almost no difference to you whether the last digit is 1 or 9. What you care about is what the highest digit is and how many digits there are in total. It is always useful if you can find all the digits. However, these problems cannot be solved at all without the use of computers. Dedicated computer software tools are also essential. If computer mathematics languages are used, for instance, with the command in MATLAB

```
>> a=sym(2019)^2019
```

all the digits in the result can be obtained as

$$a = 11\,755\,143\,827\,466\,954\,841\,312 \cdots 732\,793\,278\,321\,979.$$

There are 6 674 digits in total, which may fill about two pages of the book.

**Example 1.2.** In conventional calculus courses, the readers may have learnt how to find the derivative of a given function, with manual formulation approach. Emphasis is on manual computation in conventional courses. Assuming that a function $f(x)$ is given as $f(x) = \cos x/(x^2 + 3x + 2)$, are you able to derive manually $\mathrm{d}^4 f(x)/\mathrm{d}x^4$?

**Solutions.** The reader may use the methods learnt in calculus courses to compute the derivatives. The first-order derivative $\mathrm{d}f(t)/\mathrm{d}x$ should be evaluated first, and then the subsequent derivatives such as second-, third- and fourth-order derivatives of the given function $f(x)$ can be derived in turn. Even higher-order derivatives can be formulated in this way. The procedure is, in fact, more suitable for computers. With MATLAB, the fourth-order derivative of the function $f(x)$ can be calculated using a single line

```
>> syms x; f=cos(x)/(x^2+3*x+2); y=diff(f,x,4) % find the derivative
```

and the result obtained is

$$\begin{aligned}
\frac{\mathrm{d}^4 f(t)}{\mathrm{d}x^4} = {} & \frac{\cos x}{x^2 + 3x + 2} + \frac{12\cos x}{(x^2 + 3x + 2)^2} + \frac{24\cos x}{(x^2 + 3x + 2)^3} \\
& - \frac{12\cos x(2x + 3)^2}{(x^2 + 3x + 2)^3} - \frac{48\cos x(2x + 3)^2}{(x^2 + 3x + 2)^4} + \frac{24\cos x(2x + 3)^4}{(x^2 + 3x + 2)^5} \\
& + \frac{24\sin x(2x + 3)^3}{(x^2 + 3x + 2)^4} - \frac{4\sin x(2x + 3)}{(x^2 + 3x + 2)^2} - \frac{16\sin x(2x + 3)}{(x^2 + 3x + 2)^3} \\
& - \frac{8\sin x(8x + 12)}{(x^2 + 3x + 2)^3} - \frac{6\cos x(2x + 3)(8x + 12)}{(x^2 + 3x + 2)^4}.
\end{aligned}$$

It is immediately seen from this example that manual evaluation is a laborious and tedious work. Even a slightly careless manipulation may lead to erroneous results. Well, with tools such as MATLAB, $d^{100}f(x)/dx^{100}$ can be derived within four seconds!

**Example 1.3.** How to find the determinant of an $n \times n$ matrix?

**Solutions.** The determinant of a large matrix is recommended to be evaluated with the algebraic cofactor method in linear algebra courses. For an $n \times n$ matrix, the determinant can be evaluated from determinants of $n$ matrices of size $(n-1) \times (n-1)$. The determinant of each $(n-1) \times (n-1)$ matrix, on the other hand, can be evaluated from the determinants of $n-1$ matrices of size $(n-2) \times (n-2)$. In this manner, the determinant of an $n \times n$ matrix can be found eventually as the algebraic sum of a certain number of determinants of $1 \times 1$ matrices, i. e., the scalars themselves. It can then be concluded that the exact expression of the determinant of any given square matrix exists.

Unfortunately, the above conclusion neglected the computability and feasibility issues. The computation load could be extremely heavy, requiring $(n-1)(n+1)! + n$ operations. For instance, if $n = 25$, the number of floating point operations (flops) is $9.68 \times 10^{27}$, which is equivalent to 204 years of computation on a supercomputer of 125.4 trillion ($1.254 \times 10^{17}$) flops (i. e., the fastest supercomputer in the world in 2016, Sunway TaihuLight). Although elegant and instructive, the algebraic cofactor method is not practically feasible. In genuine applications, usually the determinants of even larger-sized matrices are expected, which is evidently not feasible by directly applying the mentioned algebraic cofactor method.

Various numerical algorithms are available in numerical analysis courses. Unfortunately, under the double-precision data type, the algorithms may lead to unexpected errors for nearly all matrices. Consider now the Hilbert matrix defined as

$$
H = \begin{bmatrix}
1 & 1/2 & 1/3 & \cdots & 1/n \\
1/2 & 1/3 & 1/4 & \cdots & 1/(n+1) \\
\vdots & \vdots & \vdots & \ddots & \vdots \\
1/n & 1/(n+1) & 1/(n+2) & \cdots & 1/(2n-1)
\end{bmatrix}.
$$

If $n = 80$, a wrong determinant $\det(H) = 0$ is found if double-precision data type is adopted. If computer mathematics language such as MATLAB is used, the exact solution of the determinant of the matrix is found within 1.23 seconds as

$$
\det(H) = \frac{1}{\underbrace{990\,301\,014\,669\,934\,778\,788\,676\,784\,101 \cdots 00\,000}_{\text{a total of 3\,789 digits, with many omitted}}} \approx 1.00979 \times 10^{-3790}.
$$

The evaluation of determinants needs computer mathematics languages, and especially for special matrices, symbolic methods should be used to avoid errors. The following MATLAB statements can be used:

```
>> tic, H=sym(hilb(80)); d=det(H); toc, d
```

**Example 1.4.** Solve the following two systems of equations:

$$\begin{cases} x + y = 35, \\ 2x + 4y = 94, \end{cases} \qquad \begin{cases} x + 3y^3 + 2z^2 = 1/2, \\ x^2 + 3y + z^3 = 2, \\ x^3 + 2z + 2y^2 = 2/4. \end{cases}$$

**Solutions.** It can be seen that the first is the well-known chicken–rabbit cage problem in Chinese legends. Even if computers are not used, the solution can easily be found. With MATLAB, the following statements can be used:

```
>> syms x y; [x0,y0]=vpasolve(x+y==35,2*x+4*y==94)
```

With powerful computer tools such as MATLAB, the solution of the second equation is as simple as the former when solving the chicken–rabbit cage problem, from the user's viewpoint. The user can represent the equations directly in symbolic expressions. Then with the same vpasolve() function call, all the 27 solutions can be found directly. When the roots are substituted back to the original equations, the error norm is as small as $10^{-34}$.

```
>> syms x y z;   % declare symbolic variables
   f1(x,y,z)=x+3*y^3+2*z^2-1/2; f2(x,y,z)=x^2+3*y+z^3-2;
   f3(x,y,z)=x^3+2*z+2*y^2-2/4; [x0,y0,z0]=vpasolve(f1,f2,f3),
   size(x0), norm([f1(x0,y0,z0) f2(x0,y0,z0) f3(x0,y0,z0)])
```

**Example 1.5.** Solve the linear programming problem given below

$$\min \qquad (-2x_1 - x_2 - 4x_3 - 3x_4 - x_5)$$

$$\boldsymbol{x}\ \text{s.t.} \begin{cases} 2x_2+x_3+4x_4+2x_5 \leqslant 54, \\ 3x_1+4x_2+5x_3-x_4-x_5 \leqslant 62, \\ x_1,x_2 \geqslant 0,\ x_3 \geqslant 3.32,\ x_4 \geqslant 0.678,\ x_5 \geqslant 2.57. \end{cases}$$

**Solutions.** Since the problem involves constrained optimization, the analytical unconstrained method by setting the derivatives of the objective function with respect to each decision variable $x_i$ to zero cannot be used. With linear programming tools in MATLAB, the following statements can be employed:

```
>> clear; P.f=[-2 -1 -4 -3 -1]; P.Aineq=[0 2 1 4 2; 3 4 5 -1 -1];
   P.Bineq=[54 62]; P.lb=[0;0;3.32;0.678;2.57]; P.solver='linprog';
   P.options=optimset; x=linprog(P) % linear programming problem solution
```

With MATLAB, the solution can be obtained directly as $x_1 = 19.7850$, $x_2 = 0$, $x_3 = 3.3200$, $x_4 = 11.3850$, and $x_5 = 2.5700$.

With the algorithms in numerical analysis or optimization courses, conventional constrained optimization problems can be solved. However, if other special constraints are introduced, for instance, the decision variables are constrained to be integers, the integer programming must be used. There are not so many books introducing software that can tackle the integer and mixed-integer programming problems. If MATLAB is used, the solution to this example problem is easily found as $x_1 = 19$, $x_2 = 0$, $x_3 = 4$, $x_4 = 10$, $x_5 = 5$.

**Example 1.6.** In many other courses, such as on complex-valued functions, integral transforms, data interpolation and fitting, partial differential equations, probability and statistics, can you still solve typical problems after the final exams?

**Example 1.7.** With the rapid development of modern science and technology, many new mathematical theories, such as those of fuzzy and rough sets, artificial neural networks, and evolutionary computing algorithms have emerged. It would be a hard and time consuming task to use such theories to solve particular problems, without using specific computer tools. If low-level programming is expected, the researcher needs to fully understand the technical contents of the theories, as well as how to implement the algorithms with computer languages. However, if the existing tools and frameworks are used instead, the problems can be solved in a much simpler manner.

In many courses, such as on electronics, electric circuits, mechanical systems, power electronics, automatic control theory, and motor drive, more sophisticated examples and problems are deliberately avoided due to the lack of high-level computer software tools support. If MATLAB or other computer mathematics languages are used extensively in those courses, complicated practical problems can be tackled. Even innovative solutions to some specific problems can be found.

### 1.1.2 Analytical and numerical solutions

The development of modern science and engineering depends heavily on mathematics. Unfortunately, the viewpoints of pure mathematicians are different from other scientists and engineers. Mathematicians are often interested in the existence and uniqueness of the solutions. They are interested in finding the analytical or closed-form solutions to mathematical problems. If solutions are not possible, they may invent special conditions such that their theoretical method can be carried out. Perfect solutions may usually be found, while unfortunately, the solutions of their revised problems may not be suitable for the original problems. Scientists and engineers are interested in finding the exact solutions to the problems, and usually do not care much about the trivial details on how the results are found, provided that the results are meaningful and reliable. If the exact solutions are not available, scientists and engineers are happy to accept approximate solutions obtained with the numerical techniques.

In real applications, it is often seen that analytical solutions do not exist. For example, the definite integral $\dfrac{2}{\sqrt{\pi}} \displaystyle\int_0^a e^{-x^2}\, dx$ is known to have no analytical solution. Mathematicians invented a special function erf($a$) to solve the problem, and do not care what the numerical value is in particular. To find an approximate solution, engineers and scientists have to adopt numerical approaches.

Another example is the irrational number $\pi$. It is known that $\pi$ has no closed-form expression. The ancient Chinese scientist and astronomer Zu Chongzhi (429–500), also known as Tsu Ch'ung-chih, found in about CE480 that the value is between 3.1415926 and 3.1415927. This value is accurate enough in most engineering and scientific applications. Even using the less accurate value of 3.14 found by Archimedes (BCE287–BCE212) in about BCE250(?), the solutions to most practical problems are often acceptable.

In many scientific and engineering areas, numerical techniques have already been widely used. For example, in aerospace and control, numerical solutions to ordinary differential equations and linear algebra problems are successfully used for centuries. In mechanical engineering, finite element methods (FEMs) have been adopted in solving partial differential equations. For simulation experiments in engineering and non-engineering areas, numerical solutions to differential and difference equations are the core techniques. In hi-tech developments, fast Fourier transform (FFT) based digital signal processing has been recognized as an essential task. There is no doubt that if one masters one or more practical computer tools, the capabilities in solving scientific computation problems will be significantly boosted, and innovative achievement can be expected.

### 1.1.3 Development of mathematical packages

The achievement in digital computer and dedicated software fueled the developments of numerical and analytical scientific computation techniques. At the early stages of numerical computation techniques, some well-established software packages played an important part. These packages include the linear algebra package LINPACK[4] and the eigenvalue-based package EISPACK[5, 16] developed in the USA, the NAG package by the Numerical Algorithm Group in the UK, and the package in the well-accepted book *Numerical Recipes*[14]. They are widely accepted by the users and with good reputation.

The famous packages such as LINPACK and EISPACK are both developed for numerical linear algebra computations. EISPACK and LINPACK packages were originally written in Fortran. An example is given next to show how to use these packages in solving linear algebra problems.

**Example 1.8.** The eigenvalues of an $N \times N$ real matrix $\boldsymbol{A}$ are expressed as $\boldsymbol{W}_{\mathrm{R}}$ and $\boldsymbol{W}_{\mathrm{I}}$ for the real and imaginary parts, respectively. The eigenvectors $\boldsymbol{Z}$ are also expected. As suggested by EISPACK, the following code is needed in the sequentially subroutine-calling sequence:

```
CALL BALANC(NM,N,A,IS1,IS2,FV1)
CALL ELMHES(NM,N,IS1,IS2,A,IV1)
CALL ELTRAN(NM,N,IS1,IS2,A,IV1,Z)
CALL HQR2(NM,N,IS1,IS2,A,WR,WI,Z,IERR)
IF (IERR.EQ.0) GOTO 99999
CALL BALBAK(NM,N,IS1,IS2,FV1,N,Z)
```

Apart from the main body of the program, the user should also write a few lines to input or initialize the matrix $\boldsymbol{A}$ to the above program and return or display the results obtained by adding some display or printing statements. Then, the whole program should be compiled and linked with the EISPACK library to generate an executable program. It can be seen that the procedure is quite complicated. Moreover, if another matrix is to be investigated, the whole procedure might be repeated, which makes the solution process even more complicated.

Unlike EISPACK and LINPACK, Numerical Recipes and NAG packages include numerical subroutines in various mathematics branches. The names of the subroutines are not quite informative, since they are composed of letters and numbers. Beginners may find it difficult to select a suitable subroutine for their own problems, not to mention the sophisticated syntaxes of the subroutines. Programming using the packages is not an easy job. Each subroutine comes with many input arguments. They are sometimes too complicated to use.

Numerical Recipes is a widely used numerical package used worldwide. There are different versions such as C, Fortran, and Pascal. It is suitable for use directly by scientific researchers and engineers. There are altogether about 200 high-efficiency and practical subroutines in the package. Generally, the subroutines are of high-standard and reliable, therefore, trusted by the researchers.

For the readers with good knowledge of languages such as Fortran and C, it is immediately found that programming for ordinary scientific computing problems is usually complicated, especially when matrices and graphics are involved. For instance, if one wants to solve a linear equation system, low-level programming is involved. The user has to write a solver, for instance, with Gaussian elimination method, then the program has to be compiled and linked. If there is no ready-to-use subroutine, the users must input the whole subroutine into a computer. For such a method, a subroutine with about 100 source code lines is expected, and it is rather a time-consuming work to validate the code and to ensure the correctness of the code. For a complicated task such as eigenvalue computation with OR method, about 500 source code lines are required. Any slight carelessness may lead to errors in the final results. These packages sometimes have limitations, since most of them may fail if complex matrices are involved.

The mathematical software packages are continuously evolving, with the leading-edge numerical algorithms being implemented. More reliable, efficient, stable, and faster packages are provided. For example, in numerical linear algebra, a new LaPACK[1] is becoming the leading package. Unlike the original purposes of the old packages, the objectives of the new ones are changed. LaPACK is no longer aiming at providing facilities or libraries for direct user applications. Instead, LaPACK provides support to professional software and languages. For example, MATLAB and a freeware Scilab[8] are not using the packages of LINPACK and EISPACK, instead LaPACK is adopted as the low-level supporting library.

Mathematical software packages appeared in certain mathematical branches so that one may call them from Fortran, C/C++, or other programming languages. Specific interface can be made so that the packages can be called from MATLAB. There are also significantly many mathematical toolboxes written in MATLAB or other computer mathematical languages. Once one has certain typical mathematical problems to be solved, the related official toolboxes in MATLAB can be called directly, since they are usually written by experts with very good reputation in the corresponding fields. The results obtained may be considered reliable, compared with those written in low-level commands.

### 1.1.4 Limitations of conventional computer languages

Many people are using conventional computer languages, such as C and Fortran, in their research. Needless to say, these languages were very useful, and they were the low-level supporting languages of the computer mathematics languages such as MATLAB. However, for the modern scientific and engineering researchers, these languages are not adequate for solving their complicated computational problems. For instance, even very experienced C programmers may not be able to write C code to find the indefinite integral of a given function, such problems involve knowledge of mathematical mechanization. Even for numerical computations, there are limitations. Here two examples are given to illustrate the problems.

**Example 1.9.** It is known that the Fibonacci sequence can be generated with the following recursive formula: $a_1 = a_2 = 1$, and $a_k = a_{k-1} + a_{k-2}$, $k = 3, 4, \ldots$ Compute its first 100 terms.

**Solutions.** Data type for each variable must be assigned first in C programming language. Since the terms in the sequence are integers, it is natural to select the data types int or long. If int is selected, the following C program can be written:

```
#include <stdio.h>
main()
{ int a1, a2, a3, i;
  a1=1; a2=1; printf("%d  %d  ",a1,a2);
  for (i=3; i<=100; i++)
  { a3=a1+a2; printf("%d  ",a3); a1=a2; a2=a3;
}}
```

The Fibonacci sequence can be obtained simply by using the program. But wait. Are the results obtained correct? If the program is executed, from the 24th term on, the value of the sequence becomes negative, and from that term on, the subsequent terms are sometimes positive, sometimes negative. It is obvious that some peculiar things must have happened in the program. The problem is caused by the int data type, since its range is $(-32767, 32767)$. If a term is beyond this range, wrong results are generated. Even if long data type is adopted instead, the correct answers may only last till about 10 more terms. To solve the problem of finding the first 100 terms, or even more, of the Fibonacci sequence is certainly beyond the capabilities of average C users, or even experienced C programmers. Extremely slight carelessness may lead to misleading results.

With the use of MATLAB, such trivial things need not be considered. The following code can be written directly:

```
>> a=[1 1];                             % the first two terms
   for i=3:100, a(i)=a(i-1)+a(i-2); end; a(end) % loop structure
```

Besides, for more accurate representation of the terms, symbolic data type can be used instead, by substituting the first statement with a=sym([1,1]). In this case, the 100th term is $a_{100} = 354\,224\,848\,179\,261\,915\,075$, and even more, the 10 000th term may be obtained, with about 32 seconds of computation, and all the 2 089 decimal digits can be found, whose display may occupy more than half a page of the book.

**Example 1.10.** Write a general purpose C program to compute the product of two matrices $A$ and $B$.

**Solutions.** If $A$ is an $n \times p$ matrix and $B$ is a $p \times m$ one, it is known from linear algebra that matrix $C$ can be obtained, whose elements are

$$c_{ij} = \sum_{k=1}^{p} a_{ik} b_{kj}, \quad i = 1, \ldots, n, j = 1, \ldots, m.$$

Based on the above algorithm, the kernel part of C program can be written in a triple-loop structure:

```
for (i=0; i<n; i++){for (j=0; j<m; j++){
  c[i][j]=0; for (k=0; k<p; k++) c[i][j]+=a[i][k]*b[k][j];}}
```

It seems again that the problem can be solved with these simple statements. Unfortunately, there is still a serious problem in the short code, the multiplicability of the two matrices is not considered. Imprecisely speaking, when the number of columns of *A* equals the number of rows of *B*, the product can be found, otherwise, they are not multiplicable. To solve the problem, an extra if statement is needed.

if cols of *A* != rows of *B*, display an error message

Unfortunately, by introducing such a statement, a new problem emerges. In mathematics, when *A* or *B* is a scalar, the product of *A* and *B* can be found, however, this case is expelled by introducing the above if statement. To solve the problem, more if statements are expected to check the scalar cases.

Even though the above modifications are made, this program is not a universal one, since complex matrices were not considered at all. More statements are needed to make the program universal.

It can be seen from the example that if C or similar computer languages are used, the programmers must be very careful to consider all the possible cases. If one or more cases were not considered, wrong or misleading results may be obtained. In MATLAB, such trivial issues need not be considered at all. The command *C=A∗B* can be used directly. If the two matrices are multiplicable, the product can be obtained, otherwise, an error message will be displayed to indicate why the product cannot be found.

## 1.2 History of computer mathematics languages

### 1.2.1 The early days of computer mathematics languages

Earlier in 1978, Professor Cleve Moler, the Chairman of the Department of Computer Science at the University of New Mexico, found that the students in his linear algebra class had difficulties in solving linear algebraic problems with the then most advanced LINPACK and EISPACK packages. He conceived and implemented MATLAB (MATrix LABoratory). The first release of MATLAB was then freely distributed. Cleve Moler and Jack Little co-founded MathWorks in 1984 to further develop the MATLAB language.

By that time, control system theory involving state space methods was rapidly developing. Significantly many algebraic problems needed to be solved. The emergence of MATLAB and its Control Systems Toolbox attracted the researchers' attention from the control community. More and more control oriented toolboxes were developed by the distinguished scholars in different control disciplines, which popularized MATLAB. It was first initiated by a numerical mathematician, but MATLAB impact and reputation were first built up in the control community and became the top-selected general purpose computer language of control scientists and engineers. With more new dedicated toolboxes appearing in many other engineering disciplines, MATLAB is becoming the *de facto* standard language in science and engineering.

The well-established Mathematica and Maple, which appeared later, are also widely used computer mathematics languages.

Besides, the free language Scilab developed by INRIA in France can also be used in solving certain scientific computing problems. The most significant benefit is that it is a free and open-source software. Unfortunately, its behavior in scientific computing has not reached the standard of the commercial languages such as MATLAB.

### 1.2.2 Representative modern computer mathematics languages

In the scientific world, there are three leading computer mathematics languages with high reputation, namely MATLAB by MathWorks, Maple by Waterloo Maple, and Mathematica by Wolfram Research. They each have their own distinguishing advantages. For example, MATLAB is easy in programming and good at numerical computation, while Mathematica and Maple are more powerful in tackling pure mathematics problems.

The numerical capability of MATLAB is much stronger. Besides, various nice toolboxes by well-known scholars can be adopted to tackle the problems with high efficiency. In addition, the symbolic computation engine in Maple was used to carry out symbolic computations, and now it is replaced by the MuPAD engine. Therefore, the symbolic computation capabilities are essentially the same as those of Mathematica and Maple for average users in engineering and science.

MATLAB is extensively used in this book series. It appears that the series is presenting some mathematical problems in certain depth. However, the ultimate objective of the series is to help the readers, after understanding roughly the mathematical background, to bypass the tedious and complex technical details of mathematics and find the reliable and accurate solutions to the mathematical problems of interest with the help of MATLAB computer mathematics language. There is no doubt that the readers' capability in tackling mathematical problems can be significantly enhanced after reading the books.

This the first volume in the series. MATLAB programming and skills are fully presented in the book. The applications of MATLAB in various branches of mathematics are covered in other volumes, where thorough explorations on related materials in other fields are studied with the tools provided in MATLAB. A MATLAB-based solution pattern is created such that much more information can be obtained from different viewpoints. Creative research can be carried out with the help of MATLAB, and the new results unknown to others may be discovered with the extensive use of the tools.

## 1.3 Three-phase solution of scientific computing problems

A three-phase methodology proposed by the author is used throughout the series in presenting mathematical problem solutions[18]. The three phases are respectively "What," "How", and "Solve." In the "What" phase, the physical explanation of the mathematical problem to be solved is presented. Even though the readers have not yet learnt the corresponding mathematics course, they can understand roughly what the problem is really about. In the "How" phase, the mathematical problem is described in a manner understandable by MATLAB. In the final "Solve" phase, appropriate MATLAB functions are called to solve the problem directly. If there is an existing MATLAB function, the syntax of the function is presented, and if not, a universal function is written to solve the problem.

**Example 1.11.** Let us revisit Example 1.5, where a linear programming problem was involved. The philosophy used in the series will be illustrated.

$$\min \quad -2x_1 - x_2 - 4x_3 - 3x_4 - x_5$$

$$\boldsymbol{x} \text{ s.t.} \begin{cases} 2x_2 + x_3 + 4x_4 + 2x_5 \leqslant 54, \\ 3x_1 + 4x_2 + 5x_3 - x_4 - x_5 \leqslant 62, \\ x_1, x_2 \geqslant 0,\ x_3 \geqslant 3.32,\ x_4 \geqslant 0.678,\ x_5 \geqslant 2.57. \end{cases}$$

**Solutions.** To solve such a problem, even though the reader may have not learnt any optimization related courses, the solution can be obtained in the following three-phases.

(1) "What" phase. In this book, we shall explain first the physical meaning of the mathematical problem. In this particular example, the mathematical formula means that under the simultaneous inequality constraints

$$\begin{cases} 2x_2 + x_3 + 4x_4 + 2x_5 \leqslant 54, \\ 3x_1 + 4x_2 + 5x_3 - x_4 - x_5 \leqslant 62, \\ x_1, x_2 \geqslant 0, \quad x_3 \geqslant 3.32, \quad x_4 \geqslant 0.678, \quad x_5 \geqslant 2.57, \end{cases}$$

we need to find a set of decision variables $x_i$ to minimize the objective function $f(\boldsymbol{x}) = -2x_1 - x_2 - 4x_3 - 3x_4 - x_5$. Even though the reader may have not studied optimization, it is not difficult to understand what he/she is expected to do from the mathematical formulas.

(2) "How" phase. We now illustrate how to represent the problem in MATLAB. The code in Example 1.5 can be used to establish variable P. The mathematical problem should be expressed in a format understandable by MATLAB.

```
>> clear; P.f=[-2 -1 -4 -3 -1];                % objective function
   P.Aineq=[0 2 1 4 2; 3 4 5 -1 -1]; P.Bineq=[54 62]; % inequalities
   P.solver='linprog'; P.lb=[0;0;3.32;0.678;2.57];    % bounds
   P.options=optimset; % the whole linear programming problem in P
```

(3) "Solve" phase. We call the solver `linprog()` directly and get the result.

```
>> x=linprog(P) % call linprog() function to solve the problem
```

**Example 1.12.** Artificial neural network is an intelligent mathematical tool widely used today. It is useful in data fitting and classification. Assume that a set of samples can be generated with

```
>> x=0:0.1:pi; y=exp(-x).*sin(2*x+2);
```

Create an artificial neural network model for the samples and fit the curve of the function.

**Solutions.** If you do not want to spend time learning theoretical aspects of artificial neural networks, and only want to use a neural network to solve the data fitting problem, consider the three-phase solution pattern presented earlier. Then a few minutes of time are needed to learn what is an artificial neural network and how it can be used. The data fitting problem can be tried with neural networks. Let us start with the three-phase solution method:

(1) What is an artificial neural network? There is no need to understand the technical details. An artificial neural network can be regarded as an information processing unit. It accepts several channels of input signals, and after internally processing the signals, the output signals can be generated.

(2) How to create a neural network in MATLAB? With function `fitnet()`, a blank neural network can be created. With the function `train()`, the parameters of the network can be obtained from the given samples, and the structure of the usable neural network is obtained as shown in Figure 1.1.

```
>> net=fitnet(5); net=train(net,x,y), view(net)
```

(3) Now we use this neural network to solve a data fitting problem. Generating a set of data to fit, the output of the neural network can be obtained. Comparisons with

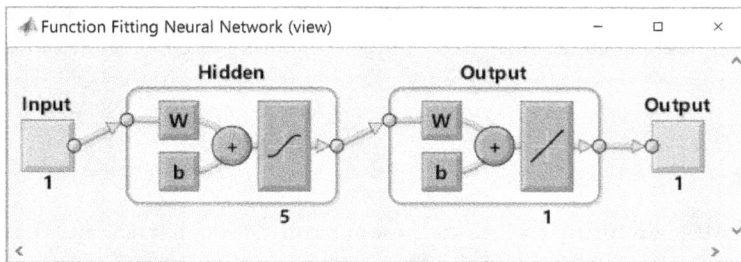

**Figure 1.1:** Structure of an artificial neural network.

the theoretical data are shown in Figure 1.2. It can be seen that even though we do not have time to learn the details of neural networks, we are still able to use them to solve our problems. Also the user may modify the structures and parameters of neural networks, for instance, the number of neurons, so as to observe the quality of data fitting through practice.

```
>> t0=0:0.01:pi; y1=net(t0); y0=exp(-t0).*sin(2*t0+2);
   plot(t0,y0,t0,y1)
```

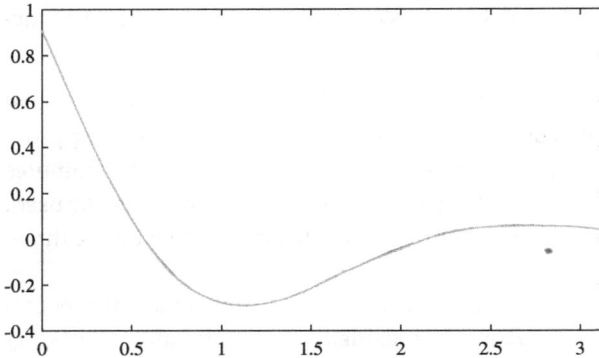

**Figure 1.2:** Fitting results with artificial neural network.

It can be seen by browsing the series that significantly many mathematical formulas are involved. Some of the knowledge seems to be rather deep in mathematics. If the readers feel unconfident about their own mathematical background, they should not worry about that. The aim of the book series is not to present technical details in mathematics, rather, the eventual goal is to help the readers first understand roughly the physical meaning of the related mathematical problems, then by avoiding the trivial details in mathematics, send the problems to computers in a way understandable by computers, and wait for the results. With the powerful tools provided by computers, our capability in dealing with practical scientific computing may well beat the first-class mathematicians who are not professionals in using computer tools. Through the study of the series, practical capabilities in scientific computing will be significantly boosted.

## 1.4 Problems

1.1    Install MATLAB environment on your machine and run the command demo. From the dialog boxes and menu items of the demonstration program, experience the powerful facilities provided in MATLAB.

1.2 Consider the example of $2019^{2019}$. Are you able to compute the exact result with C language? If not, try to solve the problem yourself in MATLAB. Hint: use sym(2019) to represent 2019.

1.3 How many digits of $\pi$ can you memorize? Try the command vpa(pi,50), and let the computer help you "memorize" the digits. Change 50 in the command to a larger number and see what happens. It is not a feasible way to tackle scientific computing by memorizing things. Even though for some people, hundreds or even thousands of digits of $\pi$ can be memorized, can you find the digits of $\sqrt{\pi}$ and $\sqrt[3]{\pi}$ by memorizing them? Try the command vpa(sym(pi)^(1/15),500) and guess what we are computing with the statements.

1.4 Consider the generalized Lyapunov equation

$$\begin{bmatrix} 8 & 1 & 6 \\ 3 & 5 & 7 \\ 4 & 9 & 2 \end{bmatrix} \boldsymbol{X} + \boldsymbol{X} \begin{bmatrix} 16 & 4 & 1 \\ 9 & 3 & 1 \\ 4 & 2 & 1 \end{bmatrix} = \begin{bmatrix} 1 & 2 & 3 \\ 4 & 5 & 6 \\ 7 & 8 & 0 \end{bmatrix}.$$

Use the command lookfor lyapunov to find a suitable solver of Lyapunov equations, related to the keyword lyapunov. Then the help command can be used to get further information, including the syntaxes of the function, so as to find the solution of the equation. Substitute the solution back to the equation and check the accuracy of the solution.

1.5 With online help facilities help sym/diff, the use of the function diff() can be found, together with the syntaxes of the function. Use online help facilities to analyze the commands provided in Example 1.2, and compare the results. If integrals are computed of the results, with repeated use of the function int(), see whether the original function can be restored.

# 2 Fundamentals of MATLAB programming

MATLAB language is the most widely used computer language in many engineering and scientific disciplines such as in automatic control and communication. Since MATLAB is used as the major computer language in the series, a comprehensive and detailed introduction is given in this book. In the series, the author concentrates on introducing MATLAB into the whole area of applied scientific computing. Good working knowledge and first-hand experience in MATLAB language will enable the readers not only understand in-depth the concepts and algorithms in research, but also increase the ability to carry out creative research work and apply MATLAB to actively tackle the problems in other related research areas.

Compared with other computer languages, MATLAB language has the following advantages:

(1) **Clarity and high efficiency.** MATLAB language is a highly integrated language. A few MATLAB sentences may carry out the work of hundreds of lines in the source code of other low-level computer languages. MATLAB programs are more reliable and easy to debug. The scientific computation efficiency is significantly boosted by using MATLAB as the major tool.

(2) **Scientific computation.** The basic element in MATLAB is a complex matrix in double-precision format. Matrix manipulations can be worked out directly. Efficient and dedicated functions for numerical computation are provided in MATLAB, such as those for solving optimization or other mathematical problems, and can be used directly. Also, symbolic computation facilities are implemented in the Symbolic Math Toolbox to support mathematical formulation.

(3) **Graphics facilities.** Another important advantage of MATLAB language is the ease in visualizing the experimental data. Implicit functions can be drawn with ease in the MATLAB environment. Moreover, the graphical user interface and object-oriented programming are also supported in MATLAB. Graphical user interface design facilities are made as simple as those provided in Visual Basic. Therefore, general purpose, user friendly programs can be written.

(4) **Comprehensive toolboxes and blocksets.** As indicated earlier, many MATLAB toolboxes and Simulink blocksets contributed by experienced programmers and trustworthy scholars are available. With the increasing popularity, MATLAB has been extended to almost all fields in science and engineering.

(5) **Powerful system simulation facilities.** The powerful block diagram-based modeling technique provided in Simulink can be adopted to analyze systems with almost any complexity. In particular, under Simulink, the control, electronic, and mechanical blocks can be modeled together under the same framework, which is currently not possible in other computer mathematics languages.

https://doi.org/10.1515/9783110666953-002

In Section 2.1, the fundamentals of MATLAB programming language are presented, including naming regulations, MATLAB reserved constants, setting of display formats, workspace setting and workspace management. In Sections 2.2–2.4, commonly used data types are presented, including double precision format, symbolic variables and functions, strings, multidimensional arrays, cells, tables, and structured variables. In Section 2.5, the fundamental statement structures are presented, including simple assignments and function calls. Also, colon expressions and submatrix extraction approaches are addressed. In Section 2.6, exchange of workspace variables and files is presented, including reading and writing of text, binary, and Microsoft Excel files.

If one masters the essential concepts and programming methodology in MATLAB, the capabilities of understanding and solving scientific computing problems may be boosted. Also, having MATLAB computation in mind may be useful in reviewing other related fields and subjects, since you have one more viewpoint of tackling problems. Sometimes you may reveal new things from these observations.

## 2.1 Command windows and fundamental commands

Essential knowledge in MATLAB programming is introduced in this section, including the variable naming regulations and MATLAB reserved constants. Also the topics, such as simple commands in MATLAB command window, setting of display format and working environment, as well as management of workspace, are presented.

### 2.1.1 Regulations in variable names

MATLAB variable names should begin by a letter, followed by numbers, letters, and underscores. For instance, MYvar12, MY_Var12, and MyVar12_ are valid variable names, while 12MyVar and _MyVar12 are invalid. MATLAB variable names are case sensitive, which means that Abc and ABc are two different variable names. One must be very careful in MATLAB programming.

It should also be noted that, if accidentally a variable name is selected the same as an existing function name, the function will be shadowed, and misleading results may be obtained. Therefore, to avoid the existing function names, command which can be used to check whether the target variable name is occupied or not.

Another way to check the existence of a name is with key=exist('name') command, where the entity name is checked. If the returned key is 1, it means that there exists a variable name in MATLAB workspace. If key is 2, file name.m exists in MATLAB search path. If the result is 3, the file name.dll exists; if 4, a Simulink model exists; if 5, a built-in MATLAB function name() exists; if 6, a pseudocode name.p exists; if 7, a folder name exists. If key is not zero, the cases in variable name should all be avoided.

**Example 2.1.** $\exp(x)$ function can be used to compute the exponential function $e^x$. If one accidentally used $\exp$ as a variable name, the function is shadowed. Try the following statements:

```
>> exp(1), exp(5), exp=3.1; exp(1), exp(5)
```

where >> is the MATLAB prompt, which is automatically given by the computer. Various commands can be issued under the prompt.

**Solutions.** Before the command $\exp=3.1$ is executed, $\exp()$ function works well. When this command is executed, function $\exp()$ is shadowed, and $\exp$ becomes a variable name, therefore, wrong results are obtained. To return to the normal stage, command clear $\exp$ should be used to delete the $\exp$ variable.

### 2.1.2 Reserved constants

Some names are reserved in MATLAB for certain commonly used constants. The names can, however, be assigned to other values. It is suggested that these names should not be reassigned to other values if possible.

(1) eps is the error tolerance for floating point operations. The default value is $2.2204 \times 10^{-16}$; if the absolute value of a quantity is smaller than eps, it can be regarded as 0.

(2) i or j. If i or j is not overwritten, they both represent $j = \sqrt{-1}$. However, they are often overwritten, for instance, when used as the variables in loops. If this happens, they can be restored with the $i=\mathrm{sqrt}(-1)$ or $i=1i$ commands.

(3) Inf stands for the MATLAB representation of infinity, $+\infty$. It can also be written as inf. Similarly, $-\infty$ can be written as $-$Inf. When 0 is used in the denominator, the value Inf can be generated, with a warning. This agrees with the IEEE standard. For mathematical computation, this definition has its advantages over that in C language.

(4) NaN (not a number), which is often returned by the operations 0/0, Inf/Inf, and others. Note that NaN times Inf returns NaN.

(5) pi means double precision representation of the circumference ratio $\pi$. The reserved digits are 3.141592653589793.

(6) true and false represent logical variables, logic 1 and logic 0.

(7) lasterr and lastwarn return the error and warning messages, respectively, received at the last time. They can be string variables, with empty string for no error, or warning message generated.

**Example 2.2.** If the radius of a circle is $r = 5$, compute its perimeter and area.

**Solutions.** Since the formulas for the perimeter and area of a circle are respectively $L = 2\pi r$ and $S = \pi r^2$, constant pi can be used in computation, with the results $L = 31.4159$, $S = 78.5398$.

```
>> r=5; L=2*pi*r, S=pi*r^2 % perimeter and area
```

MATLAB statements can be separated by commas, semicolons, and carriage returns. If a statement is terminated by a semicolon, the result is not displayed. For instance, if the statement $r = 5$ is terminated by a semicolon, the result is not displayed, while in the other two statements, since they are not terminated by semicolons, the results are displayed. The statement preceded by the percentage sign (%) marks the comments and it is not executed at all.

### 2.1.3 Setting of display formats

Under the default setting, the display format in MATLAB is short, which usually displays the data with four digits after the decimal point. If more digits are expected, the display format should be set to long, where 15 digits after the decimal point are displayed. The command to set the display format is format long. If short format is expected, the command format short should be used.

Apart from the two commonly used formats, the options of format command can also be set to compact, loose, and rat.

It is worth mentioning that the format command cannot alter the computation results, it can only be used to change the way they are displayed. The user can select the appropriate display format as needed.

**Example 2.3.** Display the results of Example 2.2 in a more accurate format. Also find the rational approximation of the results and assess the accuracy.

**Solutions.** The perimeter and area of the circle can be computed again, such that more accurate results can be obtained

```
>> format long, r=5; L=2*pi*r, S=pi*r^2
```

The new results are $L = 31.415926535897931$, $S = 78.539816339744831$.

If rational approximations of the results are expected, the display type should be set to rat, therefore

```
>> format rat, L, S, format short,
   e1=3550/113-2*pi*r, e2=8875/113-pi*r^2
```

The results are $L = 3\,550/113$, $S = 8\,875/113$. The errors in the rationalization process are respectively $e_1 = 2.6676 \times 10^{-6}$ and $e_2 = 6.6691 \times 10^{-6}$.

Command `get(0,'Format')` can be used to find the current display format.

If $a$ is a given variable, function `disp(a)` can be used directly to display the variable $a$ in MATLAB command window, where $a$ can have any data structure supported in MATLAB.

MATLAB command `type file_name` can be used to display a text file named `file_name`, and command `edit` can be used to launch a default text editor for editing the file.

### 2.1.4 Low-level operating system commands

Operating system commands can be executed directly in MATLAB using

`[status,results]=dos(command,parameters)`
`[status,results]=unix(command,parameters)`

If "status" is zero, then the execution is successful, and the "results" can be returned after the function call. The command `[s,a]=dos('dir','-echo')` can be tried and observed.

Apart from the commands `dos()` and `unix()`, the commands `cd` (change directory), `pwd` (show path), `delete` (delete file), `recycle` (recover deleted file) can also be used. Also, an executable file can be called from MATLAB. For instance, if file `mytest.exe` (or `*.com` file) is to be executed, the command `!mytest` can be called.

### 2.1.5 Setting of MATLAB working environment

When MATLAB is launched, a standard MATLAB interface is shown, and its toolbar is shown in Figure 2.1. Since the toolbar is too wide, the display is divided into left

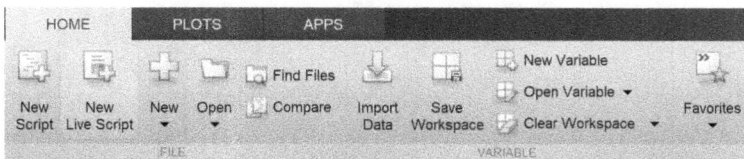

(a) left portion

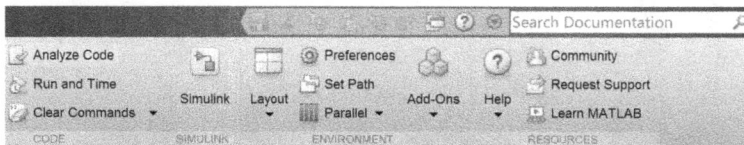

(b) right portion

**Figure 2.1:** MATLAB toolbar.

and right portion here. The file manipulating facilities are supported, and variable processing facilities are also allowed. Various simple tasks can be directly called by clicking the corresponding icons in the toolbar.

Clicking the Layout button, various MATLAB window display formats are provided. Clicking the Preferences item, the default initial setting, such as font and color, is assigned.

In real applications, you may create your own toolboxes, or download toolboxes from others. The MATLAB working path should be extended to include the paths of the new toolboxes. Clicking the Set Path icon in the toolbar, the dialog box in Figure 2.2 is shown. One may select Add Folder and Add with Subfolders, then specify the paths, and click Save button, such that when you launch MATLAB the next time, the paths are set automatically.

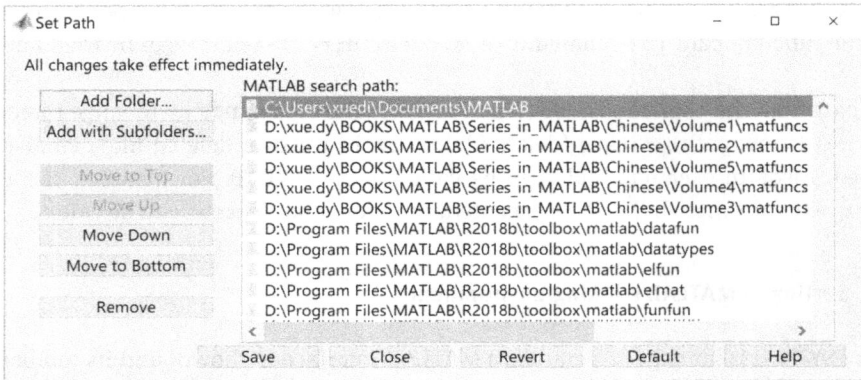

**Figure 2.2:** Path setting dialog box.

The default working folder of MATLAB is the default MATLAB root folder `bin`, and the folder is, in fact, a read-only one. It is sometimes not very convenient. It is recommended to set the MATLAB folder in the Documents. A `startup.m` file can be established, and the reader may refer to the following:

```
cd('C:\Users\xuedi\Documents\MATLAB')   % change xuedi to your own folder
format compact                          % set display format
```

Move the `startup.m` file to the `bin` folder under MATLAB's home folder. The next time MATLAB is launched, the startup file can be invoked automatically to set up the working environment.

## 2.1.6 MATLAB workspace and management

MATLAB workspace is the place where all the variables in MATLAB are stored. The command who can be used to display the variables in the current workspace, while whos can be used to show the details of the variables, such as data types and space allocations.

If one wants to remove all the variables in MATLAB workspace, the command clear command can be used. If some of the variables are to be removed, the command clear can be used, followed by the variable names, separated by spaces, to be removed. Similarly, the command clearvars can also be used to remove several variables from MATLAB workspace. The options such as -except can be used to reserve the listed variables, while removing all the other variables.

The commands save and load are a pair of commands to store and read in some variables or all the variables. Normally, the extension mat can be used, and the file is in binary format.

The command workspace can also be used to open a workspace management interface. The variables can be selected from it for further manipulation.

## 2.1.7 Other supporting facilities

In this section, some skills are presented to illustrate the facilities provided in MATLAB command window, including those for measuring elapsed time, history list, and code analyzer. The users are advised to get familiar with them and better use MATLAB for their own problems.

(1) Arrows. If the previous commands are to be issued again, the Up arrow key can be used to scroll back those commands. If one wants to issue a command started by letter a, type a in the command window, then use arrows to scroll back the previous commands.

(2) Command history window. Click the Layout button in the toolbar in Figure 2.1, from which the Command History item can be selected, and a history information window can be opened. Selecting the previously issued command from the history list, and double clicking it, the old commands can be used again. This method may sometimes be more convenient than the arrow keys.

(3) Measuring elapsed time. Two sets of time measuring commands are provided in MATLAB. One of them is to use the tic, toc command pair. Before the execution of a piece of code, the command tic should be invoked to start the stop watch, then, after execution, the toc command should be issued to read the elapsed time. The other set of commands is implemented with the cputime() function. The command $t_0$=cputime can be called before the execution of the code, to save current CPU to $t_0$, then after execution, the elapsed time can be measured with the com-

mand cputime$-t_0$. The two sets of commands are different, and can be used differently in real applications, whichever is more convenient.

(4) **Code analyzer.** Clicking **Analyze Code** button in the toolbox, an interface shown in Figure 2.3 is displayed. The files in the current MATLAB path are analyzed, and suggestions on modifications and optimizations are given. For the bk_prt.m function, two suggestions are displayed. The user can then modify the program accordingly.

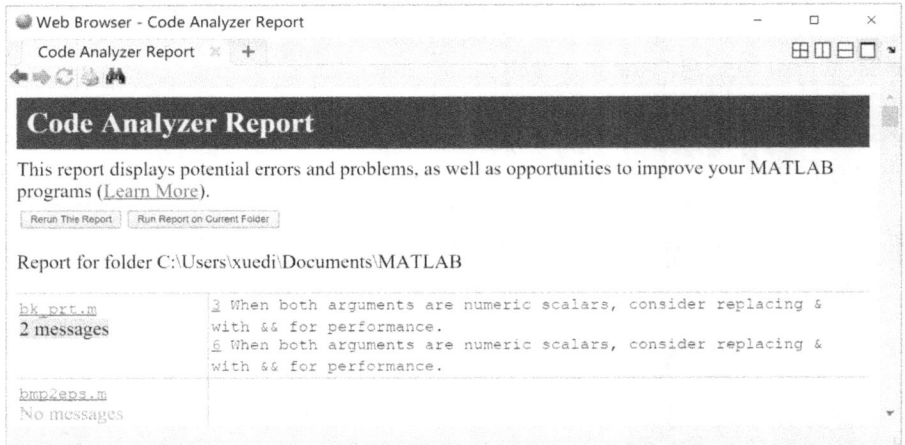

**Figure 2.3:** Code analyzer interface.

## 2.2 Commonly used data types

The most important issue in programming is data types. Two major data types commonly encountered in scientific computations are presented in this section — the numerical double precision data type, and the symbolic data type. Other data types are presented in the subsequent sections, and a solid foundation in MATLAB programming can first be established.

### 2.2.1 Numeric data types

Powerful numerical computation facilities are a prominent characteristic in MATLAB programming. In order to maintain high precision in computation, the most widely used data type is double precision floating point, with eight bytes (64 bits), obeying

IEEE standard, with 11 bits for exponent, 52 bits for storing the number, and one more bit for the sign. The range of the data is $-1.7 \times 10^{308} \sim 1.7 \times 10^{308}$, and the MATLAB function double() can be used for data type conversion.

In very special fields, single precision data type can be used occasionally. In single precision data type, 32-bit floating points are used, and normally seven effective decimal digits can be reserved. MATLAB function single() can be used in data type conversion.

The fundamental data type in MATLAB is the double precision complex matrix. The following statement can be used to input matrices in MATLAB.

**Example 2.4.** Input the following matrix into MATLAB workspace:

$$A = \begin{bmatrix} 1 & 2 & 3 \\ 4 & 5 & 6 \\ 7 & 8 & 0 \end{bmatrix}.$$

**Solutions.** It is very easy to input a matrix in MATLAB. The following MATLAB statements can be used to input the matrix directly into MATLAB workspace:

```
>> A=[1,2,3; 4 5,6; 7,8 0]    % direct input of a matrix
```

The statements can be used to input matrix in variable **A**. Meanwhile, the matrix can be displayed in MATLAB command window in the following format:

```
A =
    1    2    3
    4    5    6
    7    8    0
```

For convenient reading, the subsequent displays in MATLAB format are no longer given in the series. Instead, mathematical formats are given throughout the books.

In the input command, the elements of the matrix is specified in-between square brackets, and the elements in the same row are separated by spaces or commas, while different rows are separated by semicolons or carriage returns. With the above statement, a variable **A** is established in MATLAB workspace.

The following statement can be used to dynamically change the size of matrix **A**, by adding rows and columns, with valid MATLAB commands.

```
>> A=[[A; [1 2 3]], [1;2;3;4]] % dynamically change the matrix size
```

**Example 2.5.** Input the following complex matrix into MATLAB workspace:

$$B = \begin{bmatrix} 1+9j & 2+8j & 3+7j \\ 4+6j & 5+5j & 6+4j \\ 7+3j & 8+2j & 0+j \end{bmatrix}.$$

**Solutions.** The input of complex matrices is just as simple as that of real ones. The notations i and j are defined to denote imaginary values, such that complex quantities can be assigned in MATLAB. If j is to be specified, 1i, 1j, or sqrt(−1) should be used, rather than the direct use of i.

```
>> B=[1+9i,2+8i,3+7j; 4+6j 5+5i,6+4i; 7+3i,8+2j 1i]
```

### 2.2.2 Symbolic data

Symbolic data type is also supported in MATLAB, in contrast to numerical ones. Analytical solutions and formula derivations can be carried out based on the symbolic data type. Before using a symbolic variable, it should be declared first. The command syms can be used in variable declaration, with the syntax

syms  variable_list  variable_type

where "variable_list" is a list of variable names, separated by spaces. Commas or semicolons should not be used to avoid confusion.

If necessary, the "variable_type" should be further declared, with the options positive, integer, real, rational, and so on. If $a$ and $b$ are to be declared as symbolic variables, command such as syms $a$ $b$ should be used. The type of the symbolic variable should also be assigned, for instance, syms $a$ real can be used to defined a real symbolic variable $a$. The "variable_type" can be set as clear, indicating that the previously assigned variable type is removed, and it is restored to a conventional symbolic variable.

The variable type of symbolic variable can be extracted with assumptions() function, for instance, command syms $a$ real can be used to declare $a$ as a real symbolic variable, and assumptions($a$) can be used to return in($a$,'real').

Function $x$=symvar($f$) provided in MATLAB Symbolic Toolbox can be used to extract symbolic variables in a given symbolic expression $f$, and the list of symbolic variables are returned in vector $x$.

A variable precision arithmetic function vpa() is provided in MATLAB, and the syntaxes vpa($A$) or vpa($A,n$) can be used, where $A$ is the matrix or expression, $n$ is the number of effective digits, with the default being 32 decimal digits.

**Example 2.6.** How to describe 1/3 in MATLAB? What is $1/3 \times 0.3$ =?

**Solutions.** In conventional computer languages, double precision data type is normally supported, while symbolic data type is supported in computer mathematics languages. What are the difference between symbolic and double precision data types? The value of 1/3 cannot be stored in double precision framework, it can only be stored as 0.333333333333333, while the trailing digits are truncated. In the symbolic data framework, `sym(1/3)` can be used to store the exact value at any time, with no errors.

It is obvious that in mathematics, $1/3 \times 0.3 = 0.1$, and it is the same in the framework of symbolic variables. Now let us see what the result is under the framework of double precision scheme. It can be seen that the difference between the two quantities is $1.3878 \times 10^{-17}$, not zero.

```
>> 1/3*0.3-0.1
```

**Example 2.7.** Display the first 103 digits of the circumference ratio $\pi$.

**Solutions.** The function `vpa()` in Symbolic Toolbox can be used to display a symbolic variable in any precision. The first 103 digits of $\pi$ can be displayed with

```
>> vpa(pi,103)    % display the first 103 digits of π, more digits are possible
```

The value of $\pi$ is displayed as 3.141592653589793238462643383279502884197169399 375105820974944592307816406286208998628034825342117067982. If the number $n$ of the expected digits is not specified, the result of `vpa(pi)` becomes

$$\pi = 3.1415926535897932384626433832795.$$

It should be noted that the maximum number of digits that can be displayed with the command line here is 25 000. If more digits are expected, one may refer to the method presented later in Example 4.7.

**Example 2.8.** Compute and display the first 50 digits of the irrational number e.

**Solutions.** If the first 50 digits of e are expected, it is natural to use the command `vpa(exp(1),50)`. However, the value of e is evaluated first under the double precision framework, and then the first 50 digits are displayed, therefore, the result is imprecise. The correct way is to compute the value of e under the symbolic framework, with the command `vpa(exp(sym(1)),50)`. The result obtained is 2.7182818284590452353602874713526624977572470937.

The properties of a symbolic variable can further be set with `assume()` and `assumeAlso()` functions. For instance, if one wants to declare a real variable $x$, satisfying $-1 \leqslant x < 5$, the following MATLAB commands can be used directly:

```
>> syms x real; assume(x>=-1); assumeAlso(x<5); % set −1 ⩽ x < 5
```

The command `assumptions(x)` can be used to display the properties of $x$, with the result $[x<5,-1<=x]$.

**Example 2.9.** Declare a positive integer $k$, which is less than 3 000, and it is a multiple of 13.

**Solutions.** The value of $k_1 = \lfloor 3\,000/13 \rfloor$ should be computed first, then the following statements can be used to declare the expected positive integer $k$:

```
>> syms k1; assume(k1,'integer'); assumeAlso(k1<=floor(3000/13));
   assumeAlso(k1>0); k=13*k1
```

If there is already a variable $a$ in MATLAB workspace, then it is possible to convert it into a symbolic variable with $A$=sym($a$). Sometimes, special treatment should be applied. This is demonstrated with the following example.

**Example 2.10.** Input the value of 12345678901234567890 into MATLAB.

**Solutions.** It is quite natural to give the statement directly.

```
>> A=sym(12345678901234567890)
```

However, you may feel it bewildered, since the result is given instead by $A = 12345678901234567168$. Of course, this is not the expected one. From the mechanism in MATLAB, the value is converted to double precision first, whose first 16 digits are reliable, then converted to a symbolic one. Therefore, an error is introduced by the mechanism. The correct way to input such a long integer is to represent the whole number by a string, and convert the number into a symbolic constant with sym(). A 50-digit integer can be entered in the following statement:

```
>> B=sym('12345678901234567890123456789012345678901234567890')
```

**Example 2.11.** Convert the complex matrix in Example 2.5 into a symbolic one.

**Solutions.** The statement in Example 2.5 can be used to input the complex matrix, then function sym() can be called to complete the conversion. Compare the display formats of the two complex matrices.

```
>> B=[1+9i,2+8i,3+7j; 4+6j 5+5i,6+4i; 7+3i,8+2j 1i], B=sym(B)
```

### 2.2.3 Generation of arbitrary symbolic matrices

The function sym() in MATLAB can be used to generate an arbitrary matrix. With the syntaxes

$A$=sym('a',[$n,m$]), $B$=sym('b%d%d',[$n,m$])

arbitrary matrices $A$ and $B$ can be generated, however, the formats of the elements are different. The elements in matrix $A$ are $ai\_j$, while those in matrix $B$ are $b_{i,j}$. The latter syntax is recommended.

### 2.2.4 Symbolic functions

Symbolic functions can also be defined in a similar way with `syms()` function. The declaration of symbolic functions is demonstrated with the following example.

**Example 2.12.** Declare the symbolic functions $F(x)$ and $G(x, y, z, u)$.

**Solutions.** The independent variables should be declared as symbolic variables first, and based on them, the symbolic functions $F(x)$ and $G(x, y, z, u)$ can be declared.

```
>> syms x y z u F(x) G(x,y,z,u)
```

### 2.2.5 Integer and logic variables

In particular applications, as in the case of image processing, it is not necessary to save the information of any pixel by double precision quantities. The unsigned 8-bit integer data type is supported in MATLAB, with the `uint8()` command to convert it. The range of the data type is 0 ~ 255. This data type significantly reduces the memory needed, and the processing speed can also be increased. Furthermore, other integer data types such as `int8()`, `int16()`, `int32()`, `uint16()`, and `uint32()` are also supported.

Logic data type is also supported in MATLAB, whose values can only be selected as 0 and 1, with only one bit. The conversion function is `logical()`. Double precision or other data types can also be used as logic variables. If the value of a double precision variable is zero, it is equivalent to logic 0, otherwise, it is regarded as logic 1.

### 2.2.6 Recognition of data types

Function `key=class(a)` can be used to recognize the data type of variable $a$, while the returned variable `key` is a string. If $a$ is a symbolic variable, then `'sym'` is returned; if $a$ is a symbolic function, `'symfun'` is returned. Other supporting data types are `'double'`, `'single'`, `'int*'` (for integers, where $*$ is the number of bits), `'uint*'` (unsigned integers), `'char'` (strings), `'logical'`, `'struct'` (structured variables), `'cell'` and `'function_handle'`, and so on.

Apart from the function `class()`, a series of MATLAB functions initiated with `is` can be used to confirm the data type of a variable $a$. For instance, `key=isdouble(a)` can be used to judge whether variable $a$ has an integer data type or not. The returned variable `key` can be 1 or 0 in logic data type.

There are many similar functions, i. e., `ischar(a)` can be used to test whether $a$ is a string, `isnumeric(a)` can be used to check whether $a$ is a number. Similar to this type of functions, function isa() can also be used to confirm a certain data type. For instance, key=isa($a$,'double') can be used to check whether $a$ has a double precision data type. This function call is equivalent to the function isdouble().

### 2.2.7 Sizes and lengths of matrices

Function size() can be used to measure the size of a given matrix, while length() can be used to measure the length of a vector. The syntaxes of the functions are easy to understand

$k$=size($A$), $[n,m]$=size($A$), $n$=length($v$)

Function length() can also be used to measure the length, i. e., the maximum value of the numbers of rows and columns. Function size($A$,1) and size($A$,2) can be used respectively to measure the numbers of the rows and columns.

Function reshape() in MATLAB can be used to rearrange the elements in a matrix, with the syntaxes

$B$=reshape($A,n_1,m_1$), $B$=reshape($A,[n_1,m_1]$)

where the original $m \times n$ matrix $A$ is converted into a new $n_1 \times m_1$ matrix $B$ as long as $nm = n_1m_1$. The elements in matrix $A$ are expanded to a column vector in a column-wise manner first, then, they are sectioned into column vectors of length $n_1$. There are $m_1$ vectors which span the new matrix $B$.

MATLAB function openvar(var) opens a variable editing interface, which allows the visual edit format of a given variable, where var is a string of the variable name to be edited. A demonstration of the function will be given later.

## 2.3 String data type

String data type is also widely used in programming languages. Many input and output applications are always associated with string data type. In this section, string data type is presented, and string manipulation facilities such as string searching and string substitution are demonstrated. Finally, reading, writing, and converting strings are illustrated.

### 2.3.1 Expression of string variables

Strings are supported in MATLAB, and they are usually used to store text information. In fact, one specific application of strings was demonstrated in Example 2.10.

The representation methods of strings are different in MATLAB and other languages such as C. For instance, in MATLAB, single quotation marks are used, however, in C, double quotation marks are used. For instance, a string variable can be expressed in the following statement:

```
>> strA='Hello World!'
```

**Example 2.13.** Concatenate several strings into a bigger one. Several strings can be horizontally concatenated or manipulated through demonstrations.

**Solutions.** Chinese or other languages can also be used in a string. If there are several strings, they can be concatenated as a row vector, with the following statements:

```
>> strA='Hello World!'
   strB=' series connection of three strings. ';
   strC=[strA, strB, strA]
```

A longer string can be created, which is horizontally concatenated with the three strings.

```
'Hello World! series connection of three strings. Hello World!'
```

Several strings of different lengths can be concatenated in MATLAB into a "column vector". The function to use is str2mat().

```
>> strD=str2mat(strA,strB,strA)
```

A $3 \times 37$ string array can be established.

```
'Hello World!'
' series connection of three strings. '
'Hello World!'
```

If the first line of a string matrix is expected, the whole line can be extracted with the command strD(1,:), not strD(1), otherwise, only the first character in the line is extracted. The functions strvcat() and str2mat() are the same.

Since string variables are expressed by single quotation marks, the immediate question is how to express a single quotation mark inside a string? Two consecutive single quotation marks can be used to represent a quotation mark inside a string. Try the following statement and observe the result.

```
>> strE='In this string, single quote '' is defined.'
```

### 2.3.2 String processing methods

The following functions can be used to search, compare, and substitute strings. Meanwhile, other string-related methods are presented.

(1) String comparison. Function `strcmp()` can be used to compare two strings, with the syntax $k$=strcmp(str$_1$,str$_2$), where str$_1$ and str$_2$ are the two strings to be compared. If they are identical, the returned logic variable $k$ is 1, otherwise, it is 0.

Note that in practical MATLAB programming, the command str$_1$==str$_2$ should not be used, since if the lengths of the two strings are different, an error occurs.

(2) String search. Function `findstr()` can be used to check whether one string is a substring of the other, with the syntax $k$=findstr(str$_1$,str$_2$), where str$_1$ and str$_2$ are two strings. The returned vector $k$ stores the positions of the substring in the longer one. If the short one is not a substring, an empty matrix is returned.

**Example 2.14.** If variable strA stores the string 'Hello World!', find the positions of letter o.

**Solutions.** If letter 'o' is expected, the following commands can be used:

```
>> strA='Hello World!'; k=findstr(strA,'o')
```

and the result obtained is $k = [5, 8]$, indicating that the fifth and eighth characters are letter o. If the two input arguments in the `findstr()` function call are swapped, the result $k$ is exactly the same.

(3) String substitution. Function `strrep()` can be used to perform string substitution, with the syntax of str=strrep(str$_1$,str$_2$,str$_3$), where str$_1$ is the original string, str$_2$ is the substring to be substituted, str$_3$ is the new substring. The resulting string after substitution is returned in str. For instance, with the following statement, the new string str='HellLA WLArld!' can be obtained.

```
>> strF=strrep(strA,'o','LA')
```

(4) Measure the length of the string. Function `length()` can be used to count the number of characters in the string. For instance, $k$=length(strA) can be used, and the number of characters is $k = 12$.

(5) Delete the blanks at the end of the string. The `deblank(strA)` command can be used to delete the blanks at the end of the string strA. If one wants to delete all the blanks in string strA, the following statements can be used

str=strA(find(strA~=' ')), or str=strA(strA~=' ')

**Example 2.15.** Consider the strings in Example 2.13. Join the strings in series and count the number of the characters. Also delete all the blanks and measure the number again.

**Solutions.** The method in Example 2.13 can be used to join the substrings in series. It can be seen that the total number of characters is 61.

```
>> strA='Hello World!'; strB=' series connection of three strings. ';
   strC=[strA,strB,strA]; length(strC)  % join the string in series
   s1=strC(strC~=' '), length(s1)        % delete all the blanks and count
```

The 'HelloWorld!seriesconnectionofthreestrings.HelloWorld!' string with all the blanks deleted can be obtained, and the length is 53.

### 2.3.3 Conversion of string variables

**(1) Conversion between strings and double precision arrays.** Function double() can be used to convert the characters in the string into ASCII code, in the format of double precision vector. Function char() can be used to convert the ASCII vector back to the original string.

**Example 2.16.** Convert the string 'Hello World!' into ASCII code.

**Solutions.** The string can be entered into MATLAB workspace first, function double() can then be used for the conversion

```
>> strA='Hello World!'; v=double(strA), s1=char(v)
```

The ASCII code obtained is $v = [72, 101, 108, 108, 111, 32, 87, 111, 114, 108, 100]$, where each number corresponds to a character. If function char() is called for the result, the original string can be restored.

(2) **Conversion from symbolic expression to strings.** If a symbolic expression is given in variable $a$, function char() can be used to convert it into a string, with str=char($a$).

(3) **Convert a MATLAB variable into a string.** Several conversion functions are provided in MATLAB. If $v$ is a row vector, functions str=num2str($v$) or str=num2str($v$,$n$) can be used to convert it into a string. The default format is that four digits after the decimal point are retained. Besides, the users are allowed to assign $n$ for the number of digits.

**Example 2.17.** Observe the term 1/3 under double precision scheme, and see what is the exact value in MATLAB.

**Solutions.** To observe the data, it can be converted first into a string, to display a few more digits. It can be seen from the following statements that 1/3 in double precision scheme can be stored as 0.333333333333333314829616256247, and the first 15 digits are valid, while the other digits are irregularly generated ones.

```
>> a=1/3; num2str(a,30)
```

If $v$ is a matrix, num2str() function can be used to convert each row into a string, then form the string matrix. Apart from the function, function int2str() can be used to convert an integer vector into a string. If $v$ is not an integer vector, the elements in the vector are rounded automatically, then converted into strings.

(4) **Writing strings with specified formats.** A low-level function sprintf() is provided in MATLAB and it can be used to write strings in the specified format. The syntax of the function is similar to that in C language.

```
str=sprintf(format,a₁,a₂,...,aₘ)
```

where "format" is read/write format controls, with '%d' for integers, '%f' for floating numbers, and '%s' for characters. Therefore, $a_i$ variables can be written in the string str with user specified format.

**Example 2.18.** Use readable format to display the perimeter and area of the circle in Example 2.2.

**Solutions.** The computation statements in Example 2.2 can be used again, and the results can be generated with the following statements, such that the results displayed are more informative. The statements are

```
>> r=5; L=2*pi*r; S=pi*r^2; % compute the perimeter and area, no display
   str=sprintf('The perimeter is %f, the circle area is %f',L,S)
   disp(str)
```

The obtained string, str, can be displayed with function disp(), and the final display of the result is

```
The perimeter is 31.415927, the circle area is 78.539816
```

### 2.3.4 Executions of string commands

The command described in a string can be executed with the eval() function. In practical applications, sometimes, the user usually deliberately writes the commands to be executed into a string str, then, with command eval(str), the string can be executed to get the expected results. The following example is used to demonstrate the execution of strings.

**Example 2.19.** For a given row vector $b$, the number $n$ of the components can be measured with function `length()`. Write MATLAB statements such that the elements in the vector are assigned accordingly to variables a1, a2, ..., an.

**Solutions.** Generate first a row vector $b$. A loop structure (its details will be presented later) can be used to process each component. These common structure statements are simple to implement. The difficulty now is how to define variables with name a*. The commands can be expressed in a string $s$, then, one can call the function `eval()` to assign the variables.

```
>> b=[1 3 7 5 4 2 8 9 6 4 3 2 6]; n=length(b); % generate a row vector
   for i=1:n, s=['a' int2str(i) '=b(i);']; eval(s); end
```

For the given vector $b$, it can be seen that $n = 13$. With `who` command, it can be seen that a set of new variables a1, a2, ..., a13 is generated. Observing the values of these variables, it can be seen that the expected task is completed.

Function `feval()` in MATLAB can also be used to execute a given MATLAB function, with the syntax `feval(fun,`$p_1$`,`$p_2$`,...,`$p_n$`)`, where `fun` is a known function name, and effectively the statement is equivalent to the `fun(`$p_1$`,`$p_2$`,...,`$p_n$`)` command.

### 2.3.5 Interface of MuPAD language

The symbolic engine in MATLAB is now MuPAD. A significant number of low-level functions are provided in MuPAD. These functions cannot be called directly within MATLAB. Interfaces should be written to call these functions. The typical form of the interface is

$f$=`feval(symengine,MuPAD function name,argument list)`

where "MuPAD function name" refers to the name of low-level MuPAD function, while "argument list" is the input arguments to the MuPAD function. The use of the interface is: first assign the arguments to MuPAD, and then call the low-level functions in MuPAD directly. This syntax is demonstrated through the following example.

**Example 2.20.** For a given symbolic function $f(x)$, Padé approximation can be introduced to approximate with a rational function such that $f(x) \approx N(x)/D(x)$, where $N(x)$ and $D(x)$ are polynomials of $x$. There is no symbolic Padé approximation function, while in MuPAD, a low-level function `pade()` is provided. Unfortunately, the function cannot be called directly from MATLAB. An interface should be written. It is known that the syntax of the `pade()` function in MuPAD is $F$=`pade(`$f$`,`$x$`,[`$m$`,`$n$`])`. Write an interface of the low-level function.

**Solutions.** Function programming knowledge to be used here will be given later in Chapter 5. For now it is not necessary for the readers to understand the function. It is fine if the readers can understand the last two statements. The numerator and denominator can be converted to strings with function int2str(). Then, with feval() function, the fixed format is used to pass the arguments to MuPAD. With symeigine to invoke MuPAD engine, the symbolic Padé approximation can be obtained.

```
function p=padefrac(f,varargin)
[x,n,m]=default_vals({symvar(f),2,2},varargin{:});
orders=['[' int2str(n) ',' int2str(m) ']'];
p=feval(symengine,'pade',f,x,orders);
```

With similar ideas, the users can expand the capabilities of symbolic computation, by writing practical interfaces to MuPAD.

## 2.4 Other commonly used data types

MATLAB is a general purpose programming language. Apart from the data types for scientific computing, other data types can also be used, such as structured data, multidimensional arrays, cells, and tables. Besides, classes and objects are also supported. The definition and applications of these data types are presented in this section.

### 2.4.1 Multidimensional arrays

A three-dimensional array is an immediate extension of a matrix, which is a two-dimensional array. Three-dimensional arrays can be used in the description of color images, and also in control systems and many other fields. For instance, it can be used to describe frequency domain responses of multivariate systems. In practical programming, even higher-dimensional arrays may be used.

Three-dimensional arrays can be regarded as stacked matrices. For instance, if there are matrices $A_1, A_2, \ldots, A_m$ of the same sizes, they can be stacked to form a three-dimensional array. An illustration of a three-dimensional array is shown in Figure 2.4.

In the field of digital image processing and computer vision, an image can be expressed as a matrix, and each element in the matrix can be used to describe the grayscale of a pixel. Similar to the case in Figure 2.4, three layers of matrices can be used to represent color images, with the three matrices for the components in red, green, and blue, known as primitive colors.

Multidimensional arrays of dimension higher than three cannot be illustrated in graphs. They can only be understood, based on the descriptions of three-dimensional arrays. If one wants to construct an $m$-dimensional array, function cat() can be used,

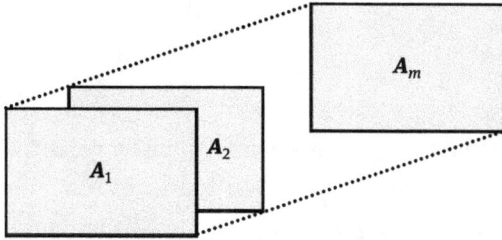

**Figure 2.4:** Illustration of a three-dimensional array.

with the syntax

$A$=cat$(m,A_1,A_2,\dots,A_m)$

where $A_1, A_2, \dots, A_m$ are $(m-1)$-dimensional arrays.

Function size() can be used to measure the size in each dimension, with size$(A,k)$ command to measure the number of elements in the $k$th dimension. If $A$ is a matrix, then $k = 1$ can be used to extract the number of rows, while $k = 2$ extracts the number of columns.

For a multidimensional array $A$, command $A(:)$ can be used to expand all the elements in $A$ into a column vector, in a column-wise manner. Function reshape() can be used to redefine the sizes of the new multidimensional array.

### 2.4.2 Cell arrays

Cells are direct extensions of matrices, and the format of the elements is stored in a similar way as in a ordinary matrix. The data types stored in each cell are not necessarily numeric quantities. Each element in the data is referred to as a "cell", which can be of any data type. For instance, $A\{i,j\}$ can be used to extract contents from the cell $A$ of the $(i,j)$th element.

**Example 2.21.** Compose a $2 \times 2$ cell array, whose four elements are of different data types.

**Solutions.** Four different variables can be entered into MATLAB first, then, they can be assigned into different cells in the cell array. The input statement is very close to that of the matrix input, and the difference is that curly brackets rather than squared ones are used.

```
>> A=[1 2 3; 4 5 6; 7 8 0]; strA='Hello World!';
   syms x; clear F; F(x)=x^2*sin(x); B={A,strA; x F}
```

The display format of the cell array is

```
2×2 cell array
  {3×3 double}      {'Hello World!'}
  {1×1 sym   }      {1×1 symfun    }
```

The elements in the cells can be extracted from $B\{i,j\}$ command. For instance, $B\{2,2\}$ can be used to extract the lower right corner element.

If the data types in each cell are identical, and the sizes are compatible, function `cell2mat()` can be used to convert the cell array into a matrix, with the syntax $A$=`cell2mat`$(C)$. Matrix $A$ can be converted to a cell array with $C$=`mat2cell`$(A,v_1, v_2)$, where submatrices can be made according to those defined by $v_1$ and $v_2$, and the converted cell array is returned in $C$.

**Example 2.22.** Consider the following partitioned matrix. Assign each submatrix into a cell to form a cell array. Also extract the whole matrix from the cell array.

$$A = \begin{bmatrix} 2 & 3 & 1 & 2 & 1 & 1 \\ 1 & 2 & 1 & 2 & 3 & 2 \\ 3 & 1 & 2 & 3 & 3 & 1 \end{bmatrix}$$

**Solutions.** The four submatrices can be input into MATLAB first, from which a cell array can be constructed. Then, `cell2mat()` function can be used to extract the whole matrix.

```
>> a11=[2 3]; a12=[1 2 1 1]; a21=[1 2; 3 1];
   a22=[1 2 3 2; 2 3 3 1]; C={a11 a12; a21 a22} % form cell array
   A=cell2mat(C)                                % extract the whole matrix
```

If the submatrix in the lower right corner cell of the cell array $C$ is expected, the command $B$=$C\{2,2\}$ can be used. However, the extracted one is in the form of a cell, which can be converted into a double precision one with $A_1$=`double`$(B)$. Or, more concisely, $A_1$=`double`$(C\{2,2\})$.

The following statements can be used to extract the whole matrix into cell arrays, whose vector $v_1$ can be selected as $v_1 = [1,2]$, meaning the first cell is with one row, and its second cell has two rows. Accordingly, $v_2 = [2,4]$, therefore, the following statements are expected:

```
>> C1=mat2cell(A,[1 2],[2 4]) % divide the matrix into partitioned cells
```

### 2.4.3 Tables

The table data type is defined in `table`, which can be used to save and handle tables or databases. Here, the data type is demonstrated through examples.

**Example 2.23.** Some of the parameters of the eight planets of the solar system are given in Table 2.1, where the relative parameters in the first four columns are the values with respect to the Earth. The unit of semimajor axis is AU (astronomical unit, defined as $149\,597\,870\,700$ m $\approx 1.5 \times 10^{11}$ m). The unit of rotation period is days. Input the table into MATLAB workspace.

**Table 2.1:** Parameters of the eight planets.

| name of planet | relative diameter | relative mass | semimajor axis | orbital period | orbital eccentricity | rotation period | confirmed moons | rings |
|---|---|---|---|---|---|---|---|---|
| Mercury | 0.382 | 0.06 | 0.39 | 0.24 | 0.206 | 58.64 | 0 | no |
| Venus | 0.949 | 0.82 | 0.72 | 0.62 | 0.007 | −243.02 | 0 | no |
| Earth | 1 | 1 | 1 | 1 | 0.017 | 1 | 1 | no |
| Mars | 0.532 | 0.11 | 1.52 | 1.88 | 0.093 | 1.03 | 2 | no |
| Jupiter | 11.209 | 317.8 | 5.20 | 11.86 | 0.048 | 0.41 | 69 | yes |
| Saturn | 9.449 | 95.2 | 9.54 | 29.46 | 0.054 | 0.43 | 62 | yes |
| Uranus | 4.007 | 14.6 | 19.22 | 84.01 | 0.047 | −0.72 | 27 | yes |
| Neptune | 3.883 | 17.2 | 30.06 | 164.8 | 0.009 | 0.67 | 14 | yes |

**Solutions.** If the entire table is to be represented in MATLAB, cells can, of course, be used. However, the `table` data type is the most suitable one, since not only the parameters of the table, but also the header of the table can be expressed. A header can be designed for this problem.

The header should be expressed in English words or letters. The following words can be expressed in strings for this example:

```
name,diameter,mass,axis,period,eccentricity,rotation,moon,ring
```

Each column of parameters can be expressed by a cell, or a column vector. Then, `table()` function can be called to construct the table data object. With the following statements, the table can be input in MATLAB workspace. The table can also be saved in the `c2dtab.mat` file:

```
>> name=str2mat('Mercury','Venus','Earth','Mars','Jupiter',...
               'Saturn','Uranus','Neptune');
   diameter=[0.382;0.949;1;0.532;11.209;9.449;4.007;3.883];
   mass=[0.06; 0.82; 1; 0.11; 317.8; 95.2; 14.6; 17.2];
   axis=[0.39; 0.72; 1; 1.52; 5.2; 9.54; 19.22; 30.06];
   period=[0.24; 0.62; 1; 1.88; 11.86; 29.46; 84.01; 164.8];
   eccentricity=[0.206; 0.007; 0.017; 0.093; 0.048; ...
       0.054; 0.047; 0.009];
   rotation=[58.64;-243.02;1;1.03;0.41;0.43;-0.72;0.67];
```

```
moon=[0; 0; 1; 2; 69; 62; 27; 14];
ring={'no';'no';'no';'no';'yes';'yes';'yes';'yes'};
planet=table(name,diameter,mass,axis,period,eccentricity,...
            rotation,moon,ring)
save c2dtab planet
```

A table variable `planet` can be established in MATLAB workspace. Since there is no semicolon used in the statement, the whole table is displayed. It can be seen that the format of the variable is straightforward. If a certain column is to be extracted, since its header term is `period`, the following statements can be used to extract the column:

```
>> planet.period
```

Function `openvar()` discussed earlier can also be used to open the interface shown in Figure 2.5. It can be seen that the `planet` variable can be edited in a visual manner.

```
>> openvar('planet')
```

planet
8x9 table

| | 1 name | 2 diameter | 3 mass | 4 axis | 5 period | 6 eccentricity | 7 rotation | 8 moon | 9 ring |
|---|---|---|---|---|---|---|---|---|---|
| 1 | Mercury | 0.3820 | 0.0600 | 0.3900 | 0.2400 | 0.2060 | 58.6400 | 0 | 'no' |
| 2 | Venus | 0.9490 | 0.8200 | 0.7200 | 0.6200 | 0.0070 | -243.0200 | 0 | 'no' |
| 3 | Earth | 1 | 1 | 1 | 1 | 0.0170 | 1 | 1 | 'no' |
| 4 | Mars | 0.5320 | 0.1100 | 1.5200 | 1.8800 | 0.0930 | 1.0300 | 2 | 'no' |
| 5 | Jupiter | 11.2090 | 317.8000 | 5.2000 | 11.8600 | 0.0480 | 0.4100 | 69 | 'yes' |
| 6 | Saturn | 9.4490 | 95.2000 | 9.5400 | 29.4600 | 0.0540 | 0.4300 | 62 | 'yes' |
| 7 | Uranus | 4.0070 | 14.6000 | 19.2200 | 84.0100 | 0.0470 | -0.7200 | 27 | 'yes' |
| 8 | Neptune | 3.8830 | 17.2000 | 30.0600 | 164.8000 | 0.0090 | 0.6700 | 14 | 'yes' |

**Figure 2.5:** Interface of `openvar()` function.

**Example 2.24.** If it is known that the mass of the Earth is $5.965 \times 10^{24}$ kg, compute the mass of the Jupiter.

**Solutions.** With the table variable `planet` in Example 2.23, the mass of the Jupiter can be evaluated directly, with $1.8957 \times 10^{27}$ kg.

```
>> load c2dtab; M=5.965e24*planet.mass(5) % load the data and compute
```

If (5) is not given in the above command, the masses of all the eight planets are computed and displayed.

A series of conversion functions related to table variables are available in MATLAB. Functions `table2cell()` and `table2struct()` are practical and can be used to convert tables to cell arrays and structured variables, respectively, while the function `table2array()` cannot be used for the variable `planet`, since it contains string items as well as numbers.

### 2.4.4 Structured variables

Structured variable, `struct`, is also suitable for describing the information in tables and databases. Low-level information, known as fields, or member variables, are used in structured variables. If a variable name is T, and one of the fields is a, the information can be accessed with T.a command. The use of structure variables is demonstrated in the following example.

**Example 2.25.** Consider again the problem in Example 2.23. Input the table to MATLAB workspace using structured data type.

**Solutions.** The input of structured variable is different from that of the table object. The following statements can be used to input the fields separately to compose the structured variable:

```
>> P.name=str2mat('Mercury','Venus','Earth','Mars','Jupiter',...
            'Saturn','Uranus','Neptune');
   P.diameter=[0.382;0.949;1;0.532;11.209;9.449;4.007;3.883];
   P.mass=[0.06; 0.82; 1; 0.11; 317.8; 95.2; 14.6; 17.2];
   P.axis=[0.39; 0.72; 1; 1.52; 5.2; 9.54; 19.22; 30.06];
   P.period=[0.24; 0.62; 1; 1.88; 11.86; 29.46; 84.01; 164.8];
   P.eccentricity=[0.206; 0.007; 0.017; 0.093; 0.048; ...
          0.054; 0.047; 0.009];
   P.rotation=[58.64;-243.02;1;1.03;0.41;0.43;-0.72;0.67];
   P.moon=[0; 0; 1; 2; 69; 62; 27; 14];
   P.ring={'no';'no';'no';'no';'yes';'yes';'yes';'yes'};
```

The contents of the structured variable can be displayed as

```
containing the following fields struct:
            name: [8×3 char]
        diameter: [8×1 double]
            mass: [8×1 double]
            axis: [8×1 double]
          period: [8×1 double]
    eccentricity: [8×1 double]
        rotation: [8×1 double]
```

```
moon: [8×1 double]
ring: {8×1 cell}
```

Alternatively, the function `struct()` can be used to input the information.

With the structure variable P, the problem in Example 2.24 can be solved directly. The result is exactly the same, and the complexity in computation is similar.

```
>> M=5.965e24*P.mass; M(5) % compute the mass of the Jupiter
```

Another set of functions to convert structured variables are also provided in MATLAB; they are `struct2cell()` and `struct2table()` for converting a structured variable into a cell or a table. Besides, function `table2array()` cannot be used to convert P into a matrix, since there are string items.

### 2.4.5 Other data types

Classes are important data types in object-oriented programming in MATLAB. Each instant of a class variable is known as an object. Variables with various complicated information can be represented in objects containing fields or member functions. Low-level operations, known as methods, are also supported. In Chapters 9 and 10, the concepts and applications of classes and objects are fully presented.

## 2.5 Fundamental statement structures

There are two fundamental statement structures, direct assignment and function call. Besides, colon expression and submatrix extraction are presented in this section.

### 2.5.1 Direct assignment statements

For direct assignment, the statement is variable=expression, where the "expression" on the right-hand side is evaluated and assigned to the "variable" in MATLAB workspace. If there is no semicolon at the end of the statement, the result of the expression is displayed in MATLAB command window. If ones does not want to display the result information, a semicolon should be appended to the statement. If the "variable" and equal sign are not given, the result is written to the reserved name ans. Therefore, ans stores the most recent information of this kind of statements.

In fact, a significant number of examples using such direct assignment statements were given earlier. In this section, we concentrate on the processing of symbolic expressions.

**Example 2.26.** If $f(x) = x^2 - x - 1$, compute $f(f(f(f(f(f(f(f(f(f(f(x))))))))))))$. If the result is a polynomial, what is its highest order?

**Solutions.** The simplest way to describe $f(x)$ is via symbolic function. The composite functions can be evaluated in a nested format directly, and the resulting polynomial can be expanded with the function `expand()`.

```
>> syms x; f(x)=x^2-x-1;
   F(x)=f(f(f(f(f(f(f(f(f(f(x))))))))))), F1=expand(F)
```

A part of the polynomial is shown below, with the highest order of 1 024:

$$F_1(x) = x^{1\,024} - 512x^{1\,023} + 130\,048x^{1\,022} - 21\,846\,272x^{1\,021} + \cdots.$$

### 2.5.2 Function call statements

Another commonly used statement format is function call, which is the mainstream MATLAB programming format. The fundamental syntax of function call is

[returned argument list]=`fun_name`(input argument list)

where `fun_name` is the function name, and the naming regulation is the same as those for variable names. Generally, the function should correspond to a *.m file in MAT-LAB search path. For instance, there should be a `my_fun.m` file for the function named `my_fun`. Of course, some functions may correspond to built-in functions in MATLAB, such as `inv()`.

The "returned argument list" and "input argument list" can be composed of several variable names, separated by commas. The returned variables can alternatively be separated by spaces as well. For instance, [$U$ $S$ $V$]=`svd`($X$) performs the singular value decomposition of the given matrix $X$, and the decomposition results are returned in the three arguments $U$, $S$, and $V$.

### 2.5.3 Functions with different syntaxes

A flexible mechanism is provided in MATLAB, which allows the user to call the same function in several different syntaxes. For instance, the built-in function `eig()` can be called with `eig`($A$), computing the eigenvalues of the given matrix; also, it can be called with [$V$,$D$]=`eig`($A$), where apart from $D$, the eigenvector matrix $V$ can also be obtained; if the function is called with `eig`($A$,$B$), generalized eigenvalue problems can be solved.

Besides, function `eig()` is provided in several different MATLAB toolboxes. In Symbolic Toolbox, function `eig()` can be used to compute analytically the eigenvalues of a symbolic matrix. In Control System Toolbox, function `eig()` can be used to compute the poles of linear systems. Since MATLAB has a good execution mechanism, the use of the functions is mutually independent. Under the executing mechanism, the data type of input argument is recognized first, then the appropriate `eig()` function is called accordingly.

The format, programming, and tactics of MATLAB functions will be fully presented in Chapter 5.

### 2.5.4 Colon expressions

Colon expressions are useful expressions in MATLAB. They are used in generating row vectors, and also can be used in submatrix extractions. The syntax of a colon expression is $v=s_1:s_2:s_3$, and a row vector $v$ is generated, with $s_1$ the starting point, $s_2$ the increment, and $s_3$ the maximum possible value of the vector. If $s_2$ and its colon are omitted, the default increment of 1 is used.

**Example 2.27.** For different increments, generate row vectors in the time interval $t \in [0, \pi]$.

**Solutions.** Taking the increment as 0.2, for example, the following statements can be used to generate a row vector:

```
>> v1=0:0.2:pi   % note that the last value is 3 rather than π
```

The row vector generated is

$$v_1 = [0, 0.2, 0.4, 0.6, 0.8, 1, 1.2, 1.4, 1.6, 1.8, 2, 2.2, 2.4, 2.6, 2.8, 3]$$

The following colon expressions can be entered:

```
>> v2=0: -0.1: pi, v3=0:pi, v4=pi:-1:0 % compare the vectors generated
```

and the generated vector $v_2$ is a $1 \times 0$ empty matrix,

$$v_3 = [0, 1, 2, 3], \quad v_4 = [3.1416, 2.1416, 1.1416, 0.1416].$$

**Example 2.28.** Find all the integers in 1~1 000, which are multiples of 13.

**Solutions.** The direct implementation is that each integer can be tested to see whether it is a multiple of 13. However, this method is rather complicated. An alternative method should be used, with the first entry of $a_1 = 13$, the second being $a_2 = a_1 + 13$, and the subsequent ones given by $a_3 = a_2 + 13$, $a_4 = a_3 + 13$, ... It is obvious that the expected row vector is started from 13, with increment of 13. Therefore, colon expression is suitable to generate the vector. The following statements in MATLAB can be used to solve the problem.

```
>> A=13: 13: 1000
```

### 2.5.5 Submatrix extractions

Submatrices can be extracted from $B=A(v_1, v_2)$, where $v_1$ is the vector containing all the rows to be extracted, while $v_2$ is the vector containing all the columns. Therefore, the submatrix $B$ can be extracted from the given matrix $A$. If $v_1$ is set to :, all the rows are extracted, while if $v_2$ is set to :, all the columns are extracted. If the keyword end is used, then the last row (or column, depending on its position) is represented.

If the $i$th row is to be deleted, the command $A(i, :)=[]$ can be used.

**Example 2.29.** With the following statements, the readers may observe from the results the actual actions of the commands:

```
>> A=[1,2,3; 4 5,6; 7,8 0];   % the semicolon at the end suppresses the display
   B1=A(1:2:end,:)            % extracts all the odd rows in A
   B2=A([3,2,1],[1,1,1])      % extracts the rows 3,2,1 in A, repeated three times
   B3=A(:,end:-1:1)           % left-right flip of A
   A(2,:)=[]; A(:,3)=[]       % deletes the second row and third column of A
```

With the above statements, the following matrices can be obtained:

$$
B_1 = \begin{bmatrix} 1 & 2 & 3 \\ 7 & 8 & 0 \end{bmatrix}, \quad
B_2 = \begin{bmatrix} 7 & 7 & 7 \\ 4 & 4 & 4 \\ 1 & 1 & 1 \end{bmatrix}, \quad
B_3 = \begin{bmatrix} 3 & 2 & 1 \\ 6 & 5 & 4 \\ 0 & 8 & 7 \end{bmatrix}, \quad
A = \begin{bmatrix} 1 & 2 \\ 7 & 8 \end{bmatrix}.
$$

**Example 2.30.** In linear algebra, the algebraic cofactor of the $(i, j)$th term in matrix $A$ can be obtained as follows: first remove the $i$th row and $j$th column, then compute the determinant of the remaining matrix, and finally, multiply the result by $(-1)^{i+j}$. Compute the $(2, 1)$th algebraic cofactor.

**Solutions.** The $(2, 1)$th algebraic cofactor can be obtained with the following statements, and the result is 24.

```
>> A=[1,2,3; 4 5,6; 7,8 0];   % the semicolon at the end suppresses the display
   i=2; j=1; B=A; B(i,:)=[]; B(:,j)=[]; d=(-1)^(i+j)*det(B)
```

### 2.5.6 Generation of equally spaced row vectors

If the increment is given, colon expressions can be used to generate an equally spaced row vector. Unfortunately, with this method, some important points may be missing. For instance, if equally spaced row vector in $[0, \pi]$ is generated, and the increment is selected as 0.1, it is natural to use the statement $v$=0:0.1:pi. However, it is found by observation that the last point is 3.1, rather than the expected $\pi$. If $\pi$ is to be reserved in the vector, the increment should be selected as $\pi/30$. Then, colon expression can be used to generate the row vector.

Two functions are implemented in MATLAB to generate different equally spaced row vectors:

(1) Linear equally spaced vectors. The function $v$=linspace($n_1$,$n_2$,$N$) can be used, where the number of points is $N$, with a default value of 50. The starting point is set to $n_1$, and the terminal value is taken as $n_2$; the generated vector is linear equally spaced.

(2) Logarithmic equally spaced vectors. In certain fields, it is usually required to have logarithmic equally spaced row vectors. For this task, the function logspace() can be used, with the syntax $w$=logspace($n_1$,$n_2$,$N$). In the statement, $N$ is the number of points, while the first and final values are respectively assigned as lg $n_1$ and lg $n_2$. Assuming that in the interval $[10^{-3}, 10^4]$, 30 points are expected, the function $w$=logspace(-3,4,30) should be specified.

## 2.6 Reading and writing of different data types

In scientific research, a significant amount of data files may be encountered. Usually, the data can be passed to other software with data files. In this section, the methods of reading and writing data or Excel files are presented.

### 2.6.1 Reading and writing of data files

As it was mentioned earlier, commands load and save can be used to read in, or write to, files. These two functions are mainly presented in this section.

The syntax of command save is

save file name variable list

where "file name" is a string, and the location is the current working folder. Also "variable list" contains the variable names to be saved, separated by spaces. Commas or other symbols cannot be used instead. If all the variables in MATLAB workspace are to be saved, it is not necessary to specify "variable list". Besides, if "file name" is not provided, default file name matlab.mat is used.

If ASCII code are expected in saving the file, the option `-ascii` is used

`save file name -ascii variable list`

If there exist spaces in the file or folder name, `load` and `save` commands may lead to errors. The functions `load()` and `save()` should be used instead.

`save(file name,'`$a_1$`','`$a_2$`',...,'`$a_m$`')`, `a=load(file name)`

where "file name" can be the name of the file, or it can be an absolute file name containing path information. Spaces are allowed in the file name. Note that, when one calls the function, the variable names to be stored should be given as strings.

**Example 2.31.** Save the table in Example 2.23 into file `my data.dat`.

**Solutions.** Since there is a space in the function name `my data.dat`, the `save` command can be used, however, misleading results may be obtained. Function `save()` should be called instead for this problem.

```
>> load c2dtab; save('my data.dat','planet')
```

### 2.6.2 Low-level reading and writing commands

Similar to C language, a set of low-level functions are provided to access files in MATLAB. The related files are:
(1) Function `fopen()` opens a file, with the syntax

   `[fid,error message]=fopen(file name,options)`

   where the available "options" are `'r'` (for read-only, as default) and `'w'` (create an empty file. If the file exists and is not empty, the file is emptied). Other options are also supported; see help for details. The returned `fid` is the file handle. If the value is $-1$, the operation is not successful. The returned "error message" is a string. Other input arguments are also supported, such as coding format, etc. Details can also be found in online help facilities.
(2) Judging whether a file is completed or not can be accomplished with `key=feof(fid)`. If the value is 1, it means that the previous file operating statement reached the end of the file, otherwise, the end of file is not reached.
(3) A line in the file can be extracted with `str=fgetl(fid)`.
(4) Low-level file operating functions, `fscanf()` and `fprintf()`, are very close to their C language prototypes. Their syntax is

   `a=fscanf(fid,format)`, `fprintf(fid,format, `$a_1,a_2,...,a_n$`)`

   where "format" is a string to control the styles, which is similar to those presented in function `sprintf()`.

(5) Function call fclose(fid) can be used to close the file. If not successful, −1 is returned.

**Example 2.32.** Write a MATLAB function with low-level commands, which can be used to display any ASCII file.

**Solutions.** The expected facility can be implemented directly with MATLAB command type. Unfortunately, the control format cannot specified, for instance, it one wants to add a blank line to each line in the ASCII file. The low-level commands will be demonstrated in the example. The first sentence is used to open an existing text file magic.m, then a loop structure can be used to read the file line by line, until the end of file, i. e., feof() function returns 1. Then the file can be closed. The information about loops and other details will be provided later.

```
>> f=fopen('magic.m') % open a file, and get the handle f
   while feof(f)~=1, disp(fgetl(f)); disp(' '), end,  fclose(f)
```

## 2.6.3 Reading and writing of Excel files

Function xlsread() in MATLAB can be used to retrieve data from a Microsoft Excel file, with

[*N*,TXT,RAW]=xlsread(file name,sheet number,range)

where "sheet number" is the number of working sheet in Excel, and "range" is a string stating the range information in the Excel format. For instance, if in the Excel file, the columns B to F and rows from 3 to 20 are to be loaded into MATLAB workspace, "range" can be set to 'B3:F20'. The numeric values in the range can be returned in matrix *N*. The text in the cells of an Excel file can be returned in variable TXT, while the original information in the Excel file can be returned in the variable RAW.

Function xlrwrite() can be used to write variables directly to Excel files from MATLAB workspace, with the syntax

xlswrite(file name,variable name,sheet number,range)

where "variable name" is the variable to write; it can be matrices and also two-dimensional cell arrays. The definitions of "sheet number" and "range" are the same as those presented earlier. If the "range" is selected as 'A1', the variable can be written to the file from the upper left corner. If it is selected as C2, the data will be saved from the second row in column C.

It is worth mentioning that, if the file is in "open" status, function xlswrite() may fail. Further error information can be given, to prompt the user to close the file first.

**Example 2.33.** Consider the table again in Example 2.23. Write it directly to an Excel file.

**Solutions.** It is pity that xlswrite() does not support direct processing of table data types. It should be converted first to a cell array. The following statements can be used, and from the upper left corner of the Excel file, an Excel file test.xls can be generated automatically, as shown in Figure 2.6. The format is exactly the same as that we are expecting. Compared with the interface in Figure 2.5, there is no table header in this file.

```
>> load c2dtab, C=table2cell(planet); % convert table to a cell array
   xlswrite('test',C,1,'A1') % last two input arguments can be omitted
```

| | A | B | C | D | E | F | G | H | I |
|---|---|---|---|---|---|---|---|---|---|
| 1 | Mercury | 0.382 | 0.06 | 0.39 | 0.24 | 0.206 | 58.64 | 0 | no |
| 2 | Venus | 0.949 | 0.82 | 0.72 | 0.62 | 0.007 | -243.02 | 0 | no |
| 3 | Earth | 1 | 1 | 1 | 1 | 0.017 | 1 | 1 | no |
| 4 | Mars | 0.532 | 0.11 | 1.52 | 1.88 | 0.093 | 1.03 | 2 | no |
| 5 | Jupiter | 11.209 | 317.8 | 5.2 | 11.86 | 0.048 | 0.41 | 69 | yes |
| 6 | Saturn | 9.449 | 95.2 | 9.54 | 29.46 | 0.054 | 0.43 | 62 | yes |
| 7 | Uranus | 4.007 | 14.6 | 19.22 | 84.01 | 0.047 | -0.72 | 27 | yes |
| 8 | Neptune | 3.883 | 17.2 | 30.06 | 164.8 | 0.009 | 0.67 | 14 | yes |

**Figure 2.6:** Screen shot of the generated Excel file.

Close the Excel file. The following statements can then be written, starting from column C, and the second row. The screen shot of the new Excel file is shown in Figure 2.7. In the overlapped part, the most recent one is updated by the new command.

```
>> xlswrite('test',C,1,'C2')   % starting from column C, second row
```

| | A | B | C | D | E | F | G | H | I | J | K |
|---|---|---|---|---|---|---|---|---|---|---|---|
| 1 | Mercury | 0.382 | 0.06 | 0.39 | 0.24 | 0.206 | 58.64 | 0 | no | | |
| 2 | Venus | 0.949 | Mercury | 0.382 | 0.06 | 0.39 | 0.24 | 0.206 | 58.64 | 0 | no |
| 3 | Earth | 1 | Venus | 0.949 | 0.82 | 0.72 | 0.62 | 0.007 | -243.02 | 0 | no |
| 4 | Mars | 0.532 | Earth | 1 | 1 | 1 | 1 | 0.017 | 1 | 1 | no |
| 5 | Jupiter | 11.209 | Mars | 0.532 | 0.11 | 1.52 | 1.88 | 0.093 | 1.03 | 2 | no |
| 6 | Saturn | 9.449 | Jupiter | 11.209 | 317.8 | 5.2 | 11.86 | 0.048 | 0.41 | 69 | yes |
| 7 | Uranus | 4.007 | Saturn | 9.449 | 95.2 | 9.54 | 29.46 | 0.054 | 0.43 | 62 | yes |
| 8 | Neptune | 3.883 | Uranus | 4.007 | 14.6 | 19.22 | 84.01 | 0.047 | -0.72 | 27 | yes |
| 9 | | | Neptune | 3.883 | 17.2 | 30.06 | 164.8 | 0.009 | 0.67 | 14 | yes |

**Figure 2.7:** The updated screen shot of the Excel file.

As a comparison, the converted structured variable obtained with the function struct2cell() in Example 2.25 cannot be written into Excel file. An alternative method can be used to convert to table data type, then to cell array. Repeating the process above, the data can be saved in the Excel file.

Considering the updated Excel file test.xls, the function xlsread() can be used to load the file

```
>> [N,txt,raw]=xlsread('test.xls')
```

and the numeric matrix can be obtained as follows. It can be seen that all the data in the field is returned in a matrix. If the corresponding matrix elements are not numbers, then NaN is automatically filled in:

$$
N = \begin{bmatrix}
0.382 & 0.06 & 0.39 & 0.24 & 0.206 & 58.64 & 0 & \text{NaN} & \text{NaN} \\
0.949 & \text{NaN} & 0.382 & 0.06 & 0.39 & 0.24 & 0.206 & 58.64 & 0 \\
1 & \text{NaN} & 0.949 & 0.82 & 0.72 & 0.62 & 0.007 & -243.02 & 0 \\
0.532 & \text{NaN} & 1 & 1 & 1 & 1 & 0.017 & 1 & 1 \\
11.209 & \text{NaN} & 0.532 & 0.11 & 1.52 & 1.88 & 0.093 & 1.03 & 2 \\
9.449 & \text{NaN} & 11.209 & 317.8 & 5.2 & 11.86 & 0.048 & 0.41 & 69 \\
4.007 & \text{NaN} & 9.449 & 95.2 & 9.54 & 29.46 & 0.054 & 0.43 & 62 \\
3.883 & \text{NaN} & 4.007 & 14.6 & 19.22 & 84.01 & 0.047 & -0.72 & 27 \\
\text{NaN} & \text{NaN} & 3.883 & 17.2 & 30.06 & 164.8 & 0.009 & 0.67 & 14
\end{bmatrix}.
$$

Other information is returned in the variable raw.

## 2.7 Problems

2.1 Invoke MATLAB environment, and write the statements

```
tic, A=rand(500); B=inv(A); norm(A*B-eye(500)), toc
```

Run the statements and observe the results. Commands help or doc can be used to check the functions you are not familiar with. Explain statement by statement the above commands.

2.2 For any integer $k$, simplify the expression $\sin(k\pi + \pi/6)$.

2.3 Predict the results of the statements $a=5$; key=isinteger($a$). What is the value of key?

2.4 For irrational numbers $\sqrt{2}$, $\sqrt[5]{11}$, $\sin 1°$, $e^2$, $\ln(21)$, and $\log_2(e)$, write the first 200 effective digits of the numbers.

2.5 Compute accurately $\lg(12345678)$. Judge which of the following statements is correct

(1) vpa(log10(sym(12345678))),    (2) vpa(sym(log10(12345678)))

2.6 Show the identities:  (1) $e^{j\pi} + 1 = 0$,  (2) $\dfrac{1 - 2\sin\alpha\cos\alpha}{\cos^2\alpha - \sin^2\alpha} = \dfrac{1 - \tan\alpha}{1 + \tan\alpha}$.

2.7　Generate a multidimensional pseudorandom array with

$A$=rand(3,4,5,6,7,8,9,10,11)

Count how many elements there are, and what the mean value is.

2.8　Input matrices $A$ and $B$ into MATLAB workspace.

$$A = \begin{bmatrix} 1 & 2 & 3 & 4 \\ 4 & 3 & 2 & 1 \\ 2 & 3 & 4 & 1 \\ 3 & 2 & 4 & 1 \end{bmatrix}, \quad B = \begin{bmatrix} 1+4j & 2+3j & 3+2j & 4+1j \\ 4+1j & 3+2j & 2+3j & 1+4j \\ 2+3j & 3+2j & 4+1j & 1+4j \\ 3+2j & 2+3j & 4+1j & 1+4j \end{bmatrix}$$

In the former $4 \times 4$ matrix, if the command $A(5,6) = 5$ is given, what will be the result?

2.9　Input the partition matrix into a $3 \times 3$ cell array.

$$A = \begin{bmatrix} 3 & 2 & 1 & 3 & 2 & 2 & 3 & 3 \\ 1 & 3 & 1 & 2 & 1 & 2 & 2 & 2 \\ 2 & 1 & 3 & 3 & 2 & 3 & 3 & 1 \\ 3 & 1 & 3 & 1 & 2 & 2 & 1 & 1 \\ 1 & 3 & 2 & 2 & 2 & 3 & 1 & 2 \end{bmatrix}$$

2.10　For the given mathematical functions $f(x) = \dfrac{x \sin x}{\sqrt{x^2 + 2(x+5)}}$, and $g(x) = \tan x$, compute the composite functions $f(g(x))$ and $g(f(x))$.

2.11　Since double precision scheme has limitations in the number of effective digits, the accuracy of the factorials of large numbers cannot be retained. Use symbolic and numerical methods to compute $C_{50}^{10}$, where, $C_m^n = m!/(n!(m-n)!)$. Function nchoosek() can be used to compute symbolically the binomial coefficients, with nchoosek(sym($m$),$n$).

2.12　List all the negative integers no less than −100, which can be divided exactly by 11. Also find all the positive integers in [3 000, 5 000], which can be divided exactly by 11.

2.13　For a given matrix $A$, write relevant MATLAB commands to extract all the even rows, and assign the result to matrix $B$. Matrix $A$=magic(8) can be used as an example to test the above statements.

2.14　Save the second to the 33th columns of a $100 \times 100$ magic matrix into an Excel file.

2.15　Find all the places of the letters 'a' and 'A' in the string 'Do you speak MAT-LAB?'. Change their cases.

2.16　Variable editing interface can be used to modify visually the matrix variable. The specific method is: Click **Open Variable** icon, click the black triangle sign, then select from the list the variable name, and then open the edit interface.

# 3 Fundamental mathematical computations

MATLAB language is one of the most widely used computer mathematics languages in the world. It has been applied successfully in a variety of disciplines, and has become the top-ranked computer language in many scientific and engineering categories.

In this chapter, essential mathematical computations in MATLAB are presented first. In Section 3.1, basic arithmetic operations with matrices are summarized, including addition, subtraction, and multiplication. Transposes and powers of matrices are also presented. The processing of complex matrices in symbolic computation is also presented. In Section 3.2, logic operations and relational functions are introduced, and a group of property judgement functions are presented. In Section 3.3, the concept and computation of transcendental functions are introduced, in particular, trigonometric, exponential, and logarithmic functions are presented. Numerical and analytical computations are all covered. Matrix transcendental functions are also addressed. In Section 3.4, simplification and conversion of symbolic expressions are presented, and then the processing of polynomials is also introduced. Simplification of trigonometric functions is studied. In Section 3.5, computation with data is studied, including integer rounding, rationalization, and sorting; maximum and minimum values are also considered. Statistical analysis of data is considered. Also the greatest common divisor, least common multiple, and prime number factorization are studied. Permutations and combinations are also presented.

## 3.1 Algebraic computation of matrices

Algebraic computation is a very attractive feature in MATLAB. The definition of algebraic computation is presented first in this section, then MATLAB-based algebraic computation and implementation are addressed.

**Definition 3.1.** Finitely many addition, subtraction, multiplication, division and exponentiation operations of variables are referred to as algebraic computation.

**Definition 3.2.** If matrix $A$ has $n$ rows and $m$ columns, then $A$ is referred to as an $n \times m$ matrix; if $n = m$, then $A$ is referred to as a square matrix.

### 3.1.1 Transposing, flipping and rotating matrices

When matrices are manipulated, transpositions, flips, and rotations are always encountered. These manipulation tasks can be completed with the following functions:

(1) Matrix transpose. The transpose of a matrix is mathematically denoted as $A^{\mathrm{T}}$. If $A$ is an $n \times m$ matrix, then the elements of its transpose $B$ satisfy $b_{ji} = a_{ij}$, $i = 1, \ldots, n$, $j = 1, \ldots, m$. Matrix $B$ is has size $m \times n$. If there exist complex elements in matrix $A$, the elements of an alternative transpose $B$ are defined as $b_{ji} = a_{ij}^*$, $i = 1, \ldots, n$, $j = 1, \ldots, m$,

https://doi.org/10.1515/9783110666953-003

i. e., transposition is carried out first, then for each element, the complex conjugate is evaluated. This type of transpose is also known as Hermitian transpose, denoted as $B = A^H$.

MATLAB command $B=A$' can be used to compute Hermitian transpose, while the transpose of matrix $A$ can be evaluated with $C=A$.'.

**Example 3.1.** Consider the complex matrix $B$ in Example 2.5. Compute its transpose and Hermitian transpose.

**Solutions.** Inputing matrix $B$ into MATLAB first, the two transposes can be computed directly with

```
>> B=[1+9i,2+8i,3+7j; 4+6j 5+5i,6+4i; 7+3i,8+2j 1i];
   B1=B', B2=B.' % two types of transpose
```

with the Hermitian transpose and direct transpose described as

$$B_1 = \begin{bmatrix} 1-9j & 4-6j & 7-3j \\ 2-8j & 5-5j & 8-2j \\ 3-7j & 6-4j & -1j \end{bmatrix}, \quad B_2 = \begin{bmatrix} 1+9j & 4+6j & 7+3j \\ 2+8j & 5+5j & 8+2j \\ 3+7j & 6+4j & 1j \end{bmatrix}.$$

(2) Matrix flipping. Various commands are provided in MATLAB for matrix flipping tasks. For instance, $B=$fliplr$(A)$ can be used to flip the matrix in left–right direction, i. e., $b_{ij} = a_{i,n+1-j}$. In fact, left–right flipping can be carried out with the function $B=A(:,end:-1:1)$;

Command $C=$flipud$(A)$ can be used to perform the up–down flipping of matrix $A$, and assign the result to matrix $C$, where $c_{ij} = a_{m+1-i,j}$. The function is equivalent to the command $C=A(end:-1:1,1)$.

(3) Matrix rotation. MATLAB function $D=$rot90$(A)$ can be used to rotate $A$ in the counterclockwise direction by 90°, and assign the result to $D$, i. e., $d_{ij} = a_{m+1-j,i}$. Function $E=$rot90$(A,k)$ can be used to rotate a matrix $90k°$ and assign the result to matrix $E$, where $k$ is an integer.

**Example 3.2.** For the given matrix $A$, rotate it by 90° in the clockwise direction, to get matrix $B$ given below:

$$A = \begin{bmatrix} 1 & 2 & 3 \\ 4 & 5 & 6 \\ 7 & 8 & 0 \end{bmatrix}, \quad B = \begin{bmatrix} 7 & 4 & 1 \\ 8 & 5 & 2 \\ 0 & 6 & 3 \end{bmatrix}.$$

**Solutions.** The standard rot90() function can be used to rotate a matrix in the counterclockwise direction. Therefore, there are two ways for the expected rotation. One way is calling rot90() function, and letting $k = -1$, while the other is letting $k = 3$, i. e., rotating in the counterclockwise direction by 270°. Therefore, the rotated matrices are $B_1 = B_2$, and they both are the needed matrix $B$.

```
>> A=[1 2 3; 4 5 6; 7 8 0]; B1=rot90(A,-1), B2=rot90(A,3)
```

### 3.1.2 Arithmetic operations

The following arithmetic computations of matrices are supported in MATLAB:

(1) Addition and subtraction. Assuming that there are two matrices $A$ and $B$ in MATLAB workspace, the commands $C=A+B$ and $C=A-B$ can be used directly. If the sizes of $A$ and $B$ are the same, the corresponding elements in $A$ and $B$ can be added or subtracted, and correct results can be returned in matrix $C$.

When the sizes are different, two special cases are also supported in MATLAB, defined as follows:

(a) If one of them is a scalar, it can be added to or subtracted from all the elements in the other matrix.

(b) If $A$ is an $n \times m$ matrix, $B$ is an $n \times 1$ column vector, or an $1 \times m$ row vector, an error message was given in earlier versions of MATLAB. In the new versions, however, the vector can be added to or subtract from all the columns or rows of the other matrix, to build the new matrix.

In other cases, error messages are given automatically, indicating that the sizes of the two matrices are incompatible.

**Example 3.3.** Observe the following two variables. What is the sum $A + B$?

$$A = \begin{bmatrix} 5 \\ 6 \end{bmatrix}, \quad B = \begin{bmatrix} 1 & 2 \\ 3 & 4 \end{bmatrix}$$

**Solutions.** The two variables cannot be added in mathematics, and also in the earlier versions of MATLAB, they cannot be added either. Error messages are given for incompatible matrices. In the recent versions of MATLAB, the following statements can be tried:

```
>> A=[5;6]; B=[1 2; 3 4]; C=A+B, D=B-A'
```

In real applications, a meaningful "addition" is defined in recent versions of MATLAB. Since $A$ is a column vector, it can be added directly to all the columns of matrix $B$, and the sum of the two variables obtained is given below. Besides, since $A^T$ is a row vector, $D$ matrix equals to a matrix where each row of $B$ is subtracted from the vector $A^T$:

$$C = \begin{bmatrix} 6 & 7 \\ 9 & 10 \end{bmatrix}, \quad D = \begin{bmatrix} -4 & -4 \\ -2 & -2 \end{bmatrix}.$$

(2) Matrix multiplication. For two given matrices $A$ and $B$, where the number of columns in matrix $A$ equals to the number of rows in matrix $B$, or one of them is a scalar, $A$ and $B$ matrices are compatible. If $A$ is an $n \times m$ matrix, while $B$ is an $m \times r$ matrix, then $C = AB$ is an $n \times r$ matrix, whose elements are

$$c_{ij} = \sum_{k=1}^{m} a_{ik} b_{kj}, \quad \text{where } i = 1, 2, \dots, n, \; j = 1, 2, \dots, r. \tag{3.1.1}$$

In MATLAB, the product of the two matrices can be computed from $C=A*B$, and it is not necessary to assign the sizes of matrices $A$ and $B$. If $A$ and $B$ are compatible, the product can be evaluated accurately and assigned to $C$; if they are incompatible, an error message is given, indicating the two matrices cannot be multiplied.

(3) Left division. The operator "\" can be used in MATLAB to compute left division of two matrices, with $A\backslash B$. Matrix $X$ is, in fact, the solution of linear equation $AX = B$. If $A$ is a nonsingular square matrix, then $X = A^{-1}B$. If $A$ is not a square matrix, $X=A\backslash B$ command can also be used to find the least squares solution $X$ of the equation $AX = B$.

(4) Right division. The operator "/" can be used in MATLAB to compute the right division of two matrices, which is the solution $X$ of the equation $XA = B$. If $A$ is a nonsingular square matrix, command $B/A$ is the solution $BA^{-1}$. More precisely, $B/A=(A'\backslash B')'$.

### 3.1.3 Complex matrices and transformations

MATLAB can be used to process complex matrices directly. Assume that, for a given complex matrix $Z$, the following simple functions can be used to transform the matrices:

(1)  Complex conjugate, with command $Z_1$=conj($Z$).
(2)  Real and imaginary parts extraction, with $R$=real($Z$) and $I$=imag($Z$).
(3)  Magnitude and phase representation, with the commands $A$=abs($Z$) and $P$= angle($Z$), where the unit of phase is radian.

In fact, the variable $Z$ is not restricted to matrices, it can also be used for multidimensional matrices and symbolic expressions.

**Example 3.4.** Consider the complex matrix $B$ in Example 2.5. Extract the real and imaginary part matrices of

$$B = \begin{bmatrix} 1+9j & 2+8j & 3+7j \\ 4+6j & 5+5j & 6+4j \\ 7+3j & 8+2j & j \end{bmatrix}.$$

**Solutions.** The matrices can be input into MATLAB first, then one can extract the real and imaginary part matrices by using

```
>> B=[1+9i,2+8i,3+7j; 4+6j 5+5i,6+4i; 7+3i,8+2j 1i];
   R=real(B), I=imag(B)
```

The real and imaginary parts obtained are

$$R = \begin{bmatrix} 1 & 2 & 3 \\ 4 & 5 & 6 \\ 7 & 8 & 0 \end{bmatrix}, \quad I = \begin{bmatrix} 9 & 8 & 7 \\ 6 & 5 & 4 \\ 3 & 2 & 1 \end{bmatrix}.$$

### 3.1.4 Powers and roots of matrices

Powers and roots are also important in algebraic matrix computations. In this section, the computation of power and root of matrices can be evaluated. It is worth mentioning that power and root computations are only applicable for square matrices. There is no definition of power or root for rectangular matrices. In fact, power and root can mathematically be uniformly represented as $A^x$. Further explanation on the manipulation are given below:

(1) Matrix powers. Matrix power can be expressed as $A^x$. If $x$ is a positive integer, the power expression $A^x$ is the multiplication of $A$ with itself $x$ times. If $x$ is a negative integer, then the matrix $A$ is multiplied by itself $-x$ times, then the inverse of the resulted matrix is taken. If $x$ is a fraction, denote $x = n/m$, where $n$ and $m$ are integers. In effect, matrix $A$ is multiplied by itself $n$ times, then we find its $m$th root. In MATLAB, both operations can be unified with a command $F = A^x$.

(2) Roots of a matrix. It can be seen in mathematics that a unique result can be obtained in the multiplication of $A$ by itself $n$ times. Its $m$th root, however, should have $m$ different solutions. Considering $\sqrt[3]{-1}$, one of the roots is $-1$, if this root is rotated in the complex plane by $120°$, another root is obtained. A further rotation by another $120°$ leads to the third solution. How to implement the rotation by $120°$? If the complex result is multiplied by a complex scalar $\delta = e^{2\pi j/3}$, the result is a rotated one. If the $m$th root is expected, a solution $A^{(n/m)}$ can be obtained, then the other solutions can be obtained by multiplying the result with $\delta_k = e^{2k\pi j/m}$, $k = 1, 2, \ldots, m - 1$.

**Example 3.5.** Consider the matrix $A$ in Example 2.4. Find its three cubic roots, and validate the results.

**Solutions.** With ^ operation, one of the cubic roots can be obtained

```
>> A=[1,2,3; 4,5,6; 7,8,0]; C=A^(1/3)
   e=norm(A-C^3) % compute and validate the cubic root
```

The root is obtained with the error matrix norm $e = 1.0145 \times 10^{-14}$. It can be seen that the result is accurate:

$$C = \begin{bmatrix} 0.7718 + j0.6538 & 0.4869 - j0.0159 & 0.1764 - j0.2887 \\ 0.8885 - j0.0726 & 1.4473 + j0.4794 & 0.5233 - j0.4959 \\ 0.4685 - j0.6465 & 0.6693 - j0.6748 & 1.3379 + j1.0488 \end{bmatrix}.$$

In fact, there should be three roots, however, only one of them is obtained. The other two roots can be computed by the rotation method, i. e., multiplying the result by complex scalars, $Ce^{j2\pi/3}$ and $Ce^{j4\pi/3}$.

```
>> j1=exp(sqrt(-1)*2*pi/3); A1=C*j1, A2=C*j1^2 % find the roots
   e1=norm(A-A1^3), e2=norm(A-A2^3)              % validate the solutions
```

The other roots can be obtained, with the errors of $10^{-14}$ as

$$A_1 = \begin{bmatrix} -0.9521 + j0.3415 & -0.2297 + j0.4296 & 0.1618 + j0.2971 \\ -0.3814 + j0.8058 & -1.1388 + j1.0137 & 0.1678 + j0.7011 \\ 0.3256 + j0.7289 & 0.2497 + j0.9170 & -1.5772 + j0.6343 \end{bmatrix},$$

$$A_2 = \begin{bmatrix} 0.1803 - j0.9953 & -0.2572 - j0.4137 & -0.3382 - j0.0084 \\ -0.5071 - j0.7332 & -0.3085 - j1.4931 & -0.6911 - j0.2052 \\ -0.7941 - j0.0825 & -0.9190 - j0.2422 & 0.2393 - j1.6831 \end{bmatrix}.$$

If the symbolic framework is used instead, variable precision algorithm can be obtained in finding the cubic root, and the error may be as low as $7.2211 \times 10^{-39}$, much lower than that obtained under double precision framework.

```
>> A=sym([1,2,3; 4,5,6; 7,8,0]); C=A^(sym(1/3));
   C=vpa(C); norm(C^3-A) % high precision solution
```

**Example 3.6.** The inverse matrix of matrix $A$ can be mathematically denoted as $A^{-1}$, and function inv($A$) can be used to evaluate the inverse. Compute the first negative power of the complex matrix in Example 2.5, and see whether the matrix equals to the inverse or not.

**Solutions.** In order to keep the accuracy, symbolic computation can be used to compute the matrices.

```
>> B=[1+9i,2+8i,3+7j; 4+6j 5+5i,6+4i; 7+3i,8+2j 1i];
   B=sym(B); B1=B^(-1), B2=inv(B), C=B1*B
```

It can be seen that the two matrices are identical. The results are validated with the above statements.

$$B_1 = B_2 = \begin{bmatrix} 13/18 - 5j/6 & -10/9 + j/3 & -1/9 \\ -7/9 + 2j/3 & 19/18 - j/6 & 2/9 \\ -1/9 & 2/9 & -1/9 \end{bmatrix}, \quad C = \begin{bmatrix} 1 & 0 & 0 \\ 0 & 1 & 0 \\ 0 & 0 & 1 \end{bmatrix}.$$

### 3.1.5 Dot operations

**Definition 3.3.** Consider two matrices, **A** and **B**, of the same size. Individual element product of them can be defined to get the matrix **C**, where $c_{ij} = a_{ij}b_{ij}$. Matrix **C** is also known as the Hadamard product of matrices **A** and **B**, or dot product.

The dot product of two matrices can be computed with **C=A.∗B**. It can be seen that the definitions of the two products are different.

Apart from the dot product, other dot operations are supported between two matrices, or a matrix and a scalar. For instance, dot power can be evaluated with operator "`.^`", dot divisions can be evaluated with "`./`" and "`.\`". If the two variables **A** and **B** are both matrices, they must be of the same size, or one of them must be a scalar, otherwise, an error message is given.

Dot operations are important in MATLAB programming. For instance, if the sketch of $x^5$ function is needed, an **x** vector can be generated first, then for each element in **x**, the fifth power is evaluated, i. e., vector $\{x_i^5\}$ is computed. The computation cannot be carried out with **x^5**, while **x.^5** should be evaluated instead.

**Example 3.7.** Consider matrix **A** in Example 2.4. Compute and explain the command **B=A.^A** and its results.

**Solutions.** From the definition of the dot power, the $(i, j)$th element in the new matrix **B** can be computed with $b_{ij} = a_{ij}^{a_{ij}}$. Therefore, the following statements can be used

```
>> A=[1,2,3; 4 5,6; 7,8 0]; B=A.^A % power of individual elements
```

and the following matrix can be obtained:

$$B = \begin{bmatrix} 1^1 & 2^2 & 3^3 \\ 4^4 & 5^5 & 6^6 \\ 7^7 & 8^8 & 0^0 \end{bmatrix} = \begin{bmatrix} 1 & 4 & 27 \\ 256 & 3125 & 46656 \\ 823543 & 16777216 & 1 \end{bmatrix}.$$

## 3.2 Logic and comparison operations

Logic and comparison operations of matrices are mainly presented in this section. Also, searching methods in matrices are allowed, and data type or attributes of variables can be assessed.

### 3.2.1 Logic operations with matrices

There was no dedicated logic data type in the old versions of MATLAB. If a value is zero, it is regarded as logic 0, otherwise, it is regarded as logic 1. In the new versions, the policy is still valid. Besides, logic data type is also supported. Assuming that matrices **A** and **B** are both $n \times m$, the following logic operations are defined:

(1) "And" operation. The operator & is supported in MATLAB, with **A**&**B**, for the "and" operation of two matrices **A** and **B**, which is defined as follows: if the corresponding elements are both logic 1, the resulting element is logic 1, otherwise it is logic 0.

(2) "Or" operation. MATLAB command **A**|**B** can be used to evaluate "or" operation of matrices **A** and **B**. If the corresponding elements are both logic 0, the result is logic 0, otherwise, the result is logic 1.

(3) "Not" operation. With the command ~**A**, the "not" operation of a matrix **A** can be performed. If the matrix element is logic 0, the result is logic 1, otherwise, the result is logic 0.

(4) "Exclusive or" operation. The "exclusive or" operation of matrices **A** and **B** can be evaluated from xor(**A**,**B**). If one of the elements is logic 1 and the other is logic 0, the result is logic 1, otherwise, the result is 0.

### 3.2.2 Comparisons of matrices

Some comparative relationships are defined in MATLAB, such **C**=**A**>**B**, meaning that when the corresponding element satisfies $a_{ij} > b_{ij}$, the result is $c_{ij} = 1$, otherwise, $c_{ij} = 0$. Other relationships are also supported, for instance, == represents equal, >= represents larger than or equal to, ~= represents not equal.

**Example 3.8.** Consider again the matrix **A** in Example 2.4. Find the logic expression of **A**>=5.

**Solutions.** The following statements can be used directly

```
>> A=[1,2,3; 4 5,6; 7,8 0]; k=A>=5
```

and the result is a logic matrix. If the corresponding element is larger than or equal to 5, the result is logic 1, otherwise logic 0. For ease of comparison, matrix **A** is also displayed:

$$k = \begin{bmatrix} 0 & 0 & 0 \\ 0 & 1 & 1 \\ 1 & 1 & 0 \end{bmatrix}, \quad A = \begin{bmatrix} 1 & 2 & 3 \\ 4 & 5 & 6 \\ 7 & 8 & 0 \end{bmatrix}.$$

### 3.2.3 Searching commands in matrix elements

Some special functions in finding subscripts or properties of matrices are provided in MATLAB, and they are useful in programming. For instance, find() can be used to find the subscripts of a matrix satisfying certain conditions, while any() or all() functions can be used to assess the behavior of the whole column in a matrix. These functions will be demonstrated through examples.

**Example 3.9.** Consider again the matrix $A$ in Example 2.4. Find the positions of the terms which are larger or equal to 5.

**Solutions.** If one wants to find the subscripts in matrix $A$, the expression $A$>=5 can be set as the condition. Then the following statements can be used to find the subscripts.

```
>> A=[1,2,3; 4 5,6; 7,8 0];
   i=find(A>=5)'  % find the subscripts satisfying the condition
```

The result obtained is a row vector $i$ = [3, 5, 6, 8]. It can be seen that the whole matrix can be expanded in a column vector, in a column-wise manner. If the elements in the column vector is larger or equal to 5, the indexes are returned.

The following statements can be used to find the rows and columns in the form of double subscripts

```
>> [i,j]=find(A>=5)  % find the double subscripts
```

The vectors of the double subscripts are $i$ = $[3, 2, 3, 2]^T$ and $j$ = $[1, 2, 2, 3]^T$, their combination $(i, j)$ satisfies the expected condition. Besides, functions all() and any() are also practical search functions.

```
>> a1=all(A>=5), a2=any(A>=5)   % observe and understand the vectors
```

The former command can be used to find whether all the elements in a column of matrix $A$ satisfy the condition, and if all the elements satisfy the condition, the result is logic 1, otherwise it is logic 0. In the latter command, if any of the elements in a column satisfies the condition, the result is logic 1, otherwise, the result is logic 0. Therefore, the returned vectors are $a_1$ = [0, 0, 0], $a_2$ = [1, 1, 1]. If one wants to judge whether all the elements in matrix $A$ are larger or equal to 5, the command can simply be changed to all($A$(:)>=5).

**Example 3.10.** An $n \times n$ magic matrix is composed of integers $1 \sim n^2$, placed into appropriate positions so that the sums of each row, each column, the diagonal and back diagonal are all the same. Magic matrices can be generated with the command $A$=magic($n$). If $n$ is relatively large, it might be difficult to find manually an element satisfying certain conditions. Find the element 123456 from an $1\,000 \times 1\,000$ magic matrix, and locate it in the matrix.

**Solutions.** Generating directly a $1\,000 \times 1\,000$ magic matrix first, function find() can be used locate the element 123456. The result is $n$ = 877, $m$ = 545, meaning the element is located at the 877th row and 545th column in the matrix.

```
>> A=magic(1000); [n,m]=find(A==123456)
```

**Example 3.11.** Consider the matrix $A$ in Example 2.4. Find all the elements larger or equal to 5, and substitute them with $-1$.

**Solutions.** The following statements can be used in MATLAB:

```
>> A=[1,2,3; 4 5,6; 7,8 0]; i=find(A>=5); A(i)=-1
```

Even simpler command $A(A>=5)=-1$ can be used for the task. This odd command may make beginning MATLAB users confused, however, it is understandable. Considering the conditions in parentheses, the conditions return a logic vector, where the positions satisfying the conditions are set to logic 1, the others being logic 0. The command $A(a)$ extracts the elements in matrix $A$, where $a$ contains logic 1. It is not strange to understand the above command.

### 3.2.4 Attribute judgement

A set of functions are provided in MATLAB, whose names are starting with is. For instance, function names isinf(), isfinite(), isnan(), isdouble(), iscell(), an so on. The physical meaning of the functions is straightforward. If the conditions are satisfied, the returned variable is logic 1, otherwise, logic 0 is returned. These functions can also be used jointly with function find(), for instance, find(isnan($A$)) function finds the locations of the elements in matrix $A$ whose elements are NaN.

## 3.3 Computation of transcendental functions

The concepts of transcendental functions are presented first in this section, then, we consider typical functions such as exponential, logarithmic, trigonometric and inverse trigonometric functions. MATLAB-based expressions of these functions are also presented in this section.

**Definition 3.4.** Transcendental functions are the functions which cannot be evaluated through finitely many additions, subtractions, multiplications, divisions and exponentiations. For instance, the exponential, logarithmic and trigonometric functions, and so on.

Normally, commonly used transcendental functions can be evaluated directly with appropriate functions in MATLAB. The syntaxes of the functions are unified, with $y$=fun($x$), where fun is the appropriate function name, $x$ is the independent variable, while $y$ is the function. The variable $x$ can be a scalar, vector, matrix, multidimensional array, and it can also be a symbolic variable. The data type and size of $y$ are exactly the same those of variable $x$. In fact, $y$ evaluates the transcendental functions on each element of the variable $x$, and it is similar to dot operations.

### 3.3.1 Exponentials and logarithmic functions

A general power function $a^x$ can be evaluated directly from `y=a.^x`. In particular, the exponential function $e^x$ can be evaluated with `y=exp(x)`.

**Example 3.12.** Show the well known identity $e^{j\pi} + 1 = 0$ with MATLAB.

**Solutions.** This identity is often regarded as the most beautiful mathematical formula. The combination of the irrational number $\pi$, imaginary number j, and exponential function yields −1. The question is then how to validate such an identity in MATLAB? The left-hand side of the expression can be input into MATLAB, then we can perform simplification, and the simplified result is 0, meaning the identity is valid.

```
>> exp(sym(pi)*1i)+1
```

The natural logarithmic function $\ln x$ can be evaluated with function `log()`; the common logarithmic function $\lg x$ can be evaluated with `log10()` function; while the base 2 logarithmic function, $\log_2 x$, can be evaluated with `log2()`. The general base $a$ logarithmic function, $\log_a x$, can be evaluated with `log(x)/log(a)`.

**Example 3.13.** Simplify the following mathematical expression:

$$f = \log_3 729 + \frac{4\ln e^3}{\lg 5\,000 - \lg 5}\log_2 17 - \log_2 83\,521.$$

**Solutions.** The expression should be input into MATLAB, then the computer can be used to evaluate the expression and the result can be found. Through symbolic computation in MATLAB, the simplified result obtained is $f = 6$.

```
>> f=log(729)/log(3)+4*log(exp(sym(3)))/...
       (log10(5000)-log10(5))*log2(17)-log2(83521)
```

### 3.3.2 Trigonometric functions

Sine, cosine, tangent, and cotangent functions are commonly used trigonometric functions. Besides, secant, cosecant, hyperbolic sine and hyperbolic cosine functions are often encountered. The definitions of these functions are presented in this section first.

**Definition 3.5.** The secant function is the reciprocal of cosine, i. e., $\sec x = 1/\cos x$, while the cosecant function is the reciprocal of sine, $\csc x = 1/\sin x$.

**Definition 3.6.** The hyperbolic sine function is defined as $\sinh x = (e^x - e^{-x})/2$, while the hyperbolic cosine function is defined as $\cosh x = (e^x + e^{-x})/2$.

Sine, cosine, tangent, and cotangent functions can be evaluated respectively with MATLAB functions `sin()`, `cos()`, `tan()`, and `cot()`. Secant and cosecant functions can be evaluated with `sec()` and `csc()`. Hyperbolic sine $\sinh x$ and hyperbolic cosine $\cosh x$ can be evaluated respectively with `sinh()` and `cosh()`. The unit of the input variables is radian. If angle is used instead, the variables can be converted with $x_1 = $`pi*`$x$`/180`. Alternatively, functions like `sind()` can be used.

The functions such as `sin()` are similar to dot operations, since sine functions can be evaluated for each element in the input argument.

**Example 3.14.** Simplify the trigonometric expression

$$T = \frac{4}{3}\cos\frac{7\pi}{3} + 3\tan^2\frac{11\pi}{6} - \frac{1}{2\cos^2(17\pi/4)} - \frac{1}{3}\sin^2\frac{\pi}{3}.$$

**Solutions.** The following statements can be used and the final result obtained is $T = 5/12$. Symbolic data type is recommended for this example. If one argument in the expression is a symbolic variable, the whole expression is evaluated under the symbolic framework, such that the exact result can be obtained.

```
>> T=4/3*cos(7*sym(pi)/3)+3*tan(11*pi/6)^2-1/(2*cos(17*pi/4)^2)...
     -1/3*sin(pi/3)^2
```

**Example 3.15.** Compute the trigonometric expression

$$\frac{\cos 40° + \sin 50°(1 + \sqrt{3}\tan 10°)}{\sin 70° \sqrt{1 + \cos 40°}}.$$

**Solutions.** Since the unit given here is degree, not the default radian, two methods can be used. One is to convert the units to radians with $x_1 = \pi x/180$, the other is to use functions as `sind()`. The following statements can be issued:

```
>> T1=(cosd(40)+sind(50)*(1+sqrt(3)*tand(10)))/...
       sind(70)/sqrt(1+cosd(40))
   p=sym(pi)/180;
   T2=(cos(40*p)+sin(50*p)*(1+sqrt(3)*tan(10*p)))/...
       sin(70*p)/sqrt(1+cos(40*p))
   vpa(T2-sqrt(2))
```

It is found that the result is $T_1 = 1.414213562373095$, which is very close to $\sqrt{2}$. Symbolic framework can be used, however, the result $\sqrt{2}$ cannot be reached, although with `vpa()` function, it can be seen that the result is arbitrarily close to $\sqrt{2}$.

**Example 3.16.** In elementary mathematics, it seems that $|\cos x| \leqslant 1$ always holds. Is that true?

**Solutions.** The condition for $|\cos x| \leqslant 1$ is that $x$ is real. If $x$ contains imaginary part, the following statements can be employed to compute the cosine function, with $a_1 = 3.762195691083631$, $a_2 = 2.032723007019666 - 3.051897799151800j$, $a_3 = 2.032723007019666 + 3.051897799151800j$, and all of their absolute values are larger than 1:

```
>> a1=cos(2i), a2=cos(1+2i), a3=cos(1-2i)
```

### 3.3.3 Inverse trigonometric functions

**Definition 3.7.** Inverse trigonometric functions are inverse functions of trigonometric functions. Take arcsine function as an example. If $\sin x = y$, its inverse function is referred to as the inverse sine function, denoted as $x = \sin^{-1} y$, or $x = \arcsin y$. The latter form is used throughout the book series.

Since the trigonometric function $y = \sin x$ is periodic, the inverse function is not single-valued. If $x$ is its inverse sine function, $x + 2k\pi$ are all inverse sine functions, where $k$ is any integer. In practical applications, the range of the arcsine function is restricted to the interval $[-\pi/2, \pi/2]$. Similarly, arctangent, arccotangent, and arccosecant functions are all restricted to the interval $[-\pi/2, \pi/2]$, while the ranges of arccosine and arcsecant functions are restricted to the interval $[0, \pi]$.

Inverse trigonometric functions can be evaluated by the functions beginning with letter a, for instance, asin(). Similar MATLAB functions include acos(), atan(), acot(), asec(), acsc(), asinh(), acosh(), and so on. The units of these functions are all radian. If degree is expected, the results can be converted with $y_1=180*y/\text{pi}$, or, alternatively, functions such as asind() can be used.

**Example 3.17.** Given $\tan(\alpha + \pi) = -1/3$, and

$$\tan(\alpha + \beta) = \frac{\sin 2(\pi/2 - \alpha) + 4\cos^2 \alpha}{10\cos^2 \alpha - \sin 2\alpha},$$

compute $\tan(\alpha + \beta)$ and $\tan \beta$.

**Solutions.** It can be seen from the first formula that, with arctangent function, the angle $\alpha$ can be evaluated. If it is substituted into another formula, the needed $\tan(\alpha+\beta)$ can be found, from which, after arctangent function evaluation, the angle $\alpha+\beta$ can be found. Since $\alpha$ is know, $\beta$ can be found. Further, the function $\tan \beta$ can be obtained. With the following MATLAB statements, the two tangent functions can be evaluated easily:

```
>> a=atan(-sym(1/3))-pi;
   T=(sin(2*(pi/2-a))+4*cos(a)^2)/(10*cos(a)^2-sin(2*a));
   T=simplify(T), b=atan(T)-a; T1=simplify(tan(b))
```

The results obtained are $\tan(\alpha + \beta) = T = 5/16$, $\tan \beta = T_1 = 31/43$.

It should be pointed out that since the solution of $\alpha$ in $\tan(\alpha + \pi) = -1/3$ is not unique, $\alpha$ should be generally written as $\alpha = \alpha + 2k\pi$, where $k$ is any integer, the following commands can be issued. The results obtained are the same as the ones given above.

```
>> syms k integer; a=atan(-sym(1/3))-pi+2*k*pi;
   T=(sin(2*(pi/2-a))+4*cos(a)^2)/(10*cos(a)^2-sin(2*a));
   T=simplify(T), b=atan(T)-a; T1=simplify(tan(b))
```

### 3.3.4 Transcendental functions of matrices

The above mentioned transcendental functions are similar to dot operations discussed earlier. In practical applications, transcendental functions of the whole matrices are sometimes expected. In this section, the exponential function is demonstrated first, followed by the evaluation of arbitrary matrix functions.

**Definition 3.8.** The exponential function $e^A$ is defined as an infinite matrix series

$$e^A = I + A + \frac{1}{2!}A^2 + \frac{1}{3!}A^3 + \frac{1}{4!}A^4 + \cdots. \tag{3.3.1}$$

Function expm() is provided in MATLAB for computing matrix exponential function, with $F$=expm($A$), and the symbolic overload function can also be used to evaluate matrix functions such as $e^{At}$. Apart from exponential functions, a universal function funm() can be used to compute other matrix functions, with the syntax of funm($B$,@funname). Here, matrix $B$ should not be too complicated, otherwise, solutions may not be found.

An even more powerful MATLAB solver is provided in [22]. The universal MATLAB function is funmsym(), which can, in theory, be used to evaluate matrix functions of any complexity. The function is provided in the package of the book series, so that it can be called directly. The syntax of the function is $F$=funmsym($A$,fun,$x$), where $x$ is an independent variable, and fun is the prototype of the matrix function. The following examples are given to demonstrate the use of the function.

**Example 3.18.** For the given matrix $A$, compute the exponential function $e^{At}$.

$$A = \begin{bmatrix} -1 & -1 & 1 & 0 \\ -1 & -1 & -1 & 0 \\ 1 & 2 & -1 & 1 \\ 0 & -1 & 0 & -2 \end{bmatrix}$$

**Solutions.** The following MATLAB statements can be used directly, in computing the matrix exponential function using two methods:

```
>> A=[-1,-1,1,0; -1,-1,-1,0; 1,2,-1,1; 0,-1,0,-2]; A=sym(A);
   syms t x; A1=expm(A*t), A2=funmsym(A,exp(x*t),x)
```

The results from the two methods are identical

$$e^{At} = \begin{bmatrix} e^{-t}+t+t^2/2 & e^{-t}-1+t^2/2 & t+t^2/2 & t^2/2 \\ -t-t^2/2 & 1-t^2/2 & -t-t^2/2 & -t^2/2 \\ 1-e^{-t}-t^2/2 & 1-e^{-t}+t-t^2/2 & 1-t^2/2 & t-t^2/2 \\ t^2/2 & t(t-2)/2 & t^2/2 & (t^2-2t+2)/2 \end{bmatrix} e^{-t}.$$

**Example 3.19.** Consider again matrix $A$ in Example 3.18. Compute the composite matrix function $\csc e^{A^2 \sin At^2} t$ where

$$A = \begin{bmatrix} -1 & -1 & 1 & 0 \\ -1 & -1 & -1 & 0 \\ 1 & 2 & -1 & 1 \\ 0 & -1 & 0 & -2 \end{bmatrix}.$$

**Solutions.** For the composite matrix function, it might be extremely difficult in the computation process, since funm() should be called several times. For this particular problem, the last statement may not yield any result.

```
>> A=[-1,-1,1,0; -1,-1,-1,0; 1,2,-1,1; 0,-1,0,-2]; A=sym(A);
   syms t x; F1=funm(expm(A^2*funm(A*t^2,@sin))*t,@csc)
```

Function funmsym() should be used instead for this example. If matrix $A$ in the composite matrix function is replaced by variable $x$, the prototype function should be denoted as $\csc(e^{x^2 \sin xt^2} t)$. Therefore, the following statements can be used to evaluate the composite matrix function:

```
>> syms t x; F=funmsym(A,csc(exp(x^2*sin(x*t^2))*t),x)
```

With the above statements, the result can be found, but it is too complicated to display. The upper left corner element is

$$f_{1,1}(t) = \frac{1}{\sin(te^{-4\sin 2t^2})} + \frac{t^2 e^{-2\sin t^2}(2\sin t^2 + t^2 \cos t^2)^2}{2\sin(t e^{-\sin t^2})}$$

$$+ \frac{t\cos(te^{-\sin t^2})e^{-\sin t^2}(2\sin t^2 + 4t^2\cos t^2 - t^4\sin t^2)}{2\sin^2(te^{-\sin t^2})}$$

$$+ \frac{t^2\cos^2(te^{-\sin t^2})e^{-2\sin t^2}(2\sin t^2 + t^2\cos t^2)^2}{\sin^3(te^{-\sin t^2})}$$

$$- \frac{t\cos(te^{-\sin t^2})e^{-\sin t^2}(2\sin t^2 + t^2\cos t^2)(1 + 2\sin t^2 + t^2\cos t^2)}{2\sin^2(te^{-\sin t^2})}.$$

**Example 3.20.** Consider the matrix $A$ in Example 3.18. Compute $A^k$ and $k^A$, where $k$ is any integer.

**Solutions.** If matrix $A$ is substituted by scalar variable $x$, the prototypes of the matrix functions are respectively $x^k$ and $k^x$, therefore, the following statements can be used to compute $A^k$ and $k^A$:

```
>> A=[-1,-1,1,0; -1,-1,-1,0; 1,2,-1,1; 0,-1,0,-2]; A=sym(A);
   syms t x k;  A1=funmsym(A,x^k,x), A2=funmsym(A,k^x,x)
```

where $k^A$ can be obtained as

$$A_1 = (-1)^k \begin{bmatrix} 2^k - 3k/2 + k^2/2 & 2^k - 1 + k(k-1)/2 & k(k-3)/2 & k(k-1)/2 \\ -k(k-3)/2 & (-k^2 + k + 2)/2 & -k(k-3)/2 & -k(k-1)/2 \\ 1 - 2^k - k(k-1)/2 & 1 - k/2 - 2^k - k^2/2 & (-k^2 + k + 2)/2 & -k(k+1)/2 \\ k(k-1)/2 & k(k+1)/2 & k(k-1)/2 & (k^2 + k + 2)/2 \end{bmatrix}$$

$$A_2 = \frac{1}{k} \begin{bmatrix} \ln k + \ln^2 k/2 + 1/k & \ln^2 k/2 - 1 + 1/k & \ln k + \ln^2 k/2 & \ln^2 k/2 \\ -\ln k - \ln^2 k/2 & 1 - \ln^2 k/2 & -\ln k - \ln^2 k/2 & -\ln^2 k/2 \\ 1 - \ln^2 k/2 - 1/k & \ln k - \ln^2 k/2 + 1 - 1/k & 1 - \ln^2 k/2 & \ln k - \ln^2 k/2 \\ \ln^2 k/2 & \ln^2 k/2 - \ln k & \ln^2 k/2 & \ln^2 k/2 - \ln k + 1 \end{bmatrix}.$$

It should be noted that, in the latter two examples, function `funm()` cannot be used, and one must rely on the function `funmsym()` to find the solutions.

## 3.4 Simplifications and conversions of symbolic expressions

Polynomial is a commonly used mathematical expression. In this section, input and processing of symbolic polynomials are presented. Also, some processing methods such as variable substitution and dedicated simplification functions are introduced.

### 3.4.1 Polynomial operations

Polynomials can be input into MATLAB easily. Also, some processing functions are provided. For instance, function `collect()` can be used to collect like terms in a polynomial, function `expand()` can be used to expand the polynomials, while function `factor()` can be used to factorize a given polynomial.

**Example 3.21.** For a polynomial containing factors

$$P(s) = (s + 3)^2(s^2 + 3s + 2)(s^3 + 12s^2 + 48s + 64),$$

various processing commands can be tried, and the results observed.

**Solutions.** A symbolic variable $s$ should be entered first, then we can input the polynomial

```
>> syms s; P(s)=(s+3)^2*(s^2+3*s+2)*(s^3+12*s^2+48*s+64) % input P
   P1=expand(P), P2=factor(P), P3=prod(P2)   % various conversions
```

The expanded polynomial is as follows:

$$P_1(s) = s^7 + 21s^6 + 185s^5 + 883s^4 + 2\,454s^3 + 3\,944s^2 + 3\,360s + 1\,152.$$

With function `factor()`, the factors can be obtained and returned in $P_2$ vector. Further, with function `prod()`, the factorized form of the polynomial is returned in $P_3$.

$$P_2(s) = [s + 3, s + 3, s + 2, s + 1, s + 4, s + 4, s + 4],$$
$$P_3(s) = (s + 1)(s + 2)(s + 3)^2(s + 4)^3.$$

**Example 3.22.** Taylor series expansion of function $f(x) = e^{ax}\sin(b+x)$ can be obtained with `taylor()` function, in the form of a polynomial. However, the display is not quite clearly presented. Collect the terms according to $x$.

**Solutions.** The following commands can be used to compute Taylor series, then we can collect the like terms in the results.

```
>> syms x a b; f(x)=exp(a*x)*sin(b+x); F=taylor(f,'Order',5)
   collect(F)
```

The result after term collection is given as

$$F(x) = (a^4 \sin b/24 + a^3 \cos b/6 - a^2 \sin b/4 - a \cos b/6 + \sin b/24)x^4$$
$$+ (a^3 \sin b/6 + a^2 \cos b/2 - a \sin b/2 - \cos b/6)x^3$$
$$+ (a^2 \sin b/2 + a \cos b - \sin b/2)x^2 + (\cos b + a \sin b)x + \sin b.$$

### 3.4.2 Conversions and simplifications of trigonometric functions

A great number of formulas are available for trigonometric functions. For instance, the trigonometric functions of $(x + y)$ can be expanded as trigonometric functions of $x$ and $y$.

**Theorem 3.1.** *The expansion formulas of trigonometric functions are*

$$\sin(x + y) = \cos x \sin y + \cos y \sin x, \quad \cos(x + y) = \cos x \cos y - \sin x \sin y. \quad (3.4.1)$$

With Symbolic Toolbox in MATLAB, some of the formulas can be obtained. For instance, $F_1$=simplify($F$) may be used to find the simplest form, while function expand() can be used in the trigonometric function expansions.

**Example 3.23.** With Symbolic Toolbox of MATLAB, the formula in (3.4.1) can be derived. Which function can we use to transform the formula from right to left?

**Solutions.** Formulas in (3.4.1) can be obtained directly with expand() function. With function simplify(), the simplified form on the left can be restored.

```
>> syms x y; F1=expand(sin(x+y)), F2=expand(cos(x+y))
   simplify(F1), simplify(F2), F3=expand(tan(x+y))
```

The above statement can also be used to expand $\tan(x + y)$ as

$$\tan(x + y) = \frac{\tan x + \tan y}{1 - \tan x \tan y}.$$

### 3.4.3 Simplification of symbolic expressions

Symbolic Toolbox functions can be used in deriving formulas, in particular, simplify() function can be used in results simplification. Sometimes, the simplified results are not in their simplest forms, or not in the forms we are expecting. Further simplification may be needed.

The function simplify() can be tried first, with $s_1$=simplify($s$). Various simplification functions are tried automatically to simplify the expression $s$. The simplest form of $s$ can be obtained. The simple() function in the old versions of MATLAB can no longer be used.

Apart from simplify() function, function numden() can be used to extract the numerator and denominator from an expression. The syntaxes and information of the functions can be obtained with help command.

User-defined simplification function $F_1$=rewrite($F$,fun) is provided in MATLAB, where fun is the user-specified function. For instance, 'sin' can be used to simplify the results where only sine function is retained. The options 'sincos' (where only sine and cosine functions are retained), 'cos' (only cosine function), 'tan' (only tangent function) can be used; and also the options 'exp' (only exponential function), 'sqrt' (only square root function), 'log' (only logarithmic function), 'heaviside' (only Heaviside function) and 'piecewise' (only piecewise function) can be employed.

**Example 3.24.** Expand the $\tan(x + y)$ function in Example 3.23, and convert the expression into an expression with only sine functions.

**Solutions.** The expanded form can be obtained first, and then sine or cosine function conversion can obtained directly.

```
>> syms x y, F=tan(x+y), F1=rewrite(expand(F),'sin')
```

The result obtained is

$$F_1 = \frac{\sin x/(2\sin^2 x/2 - 1) + \sin y/(2\sin^2 y/2 - 1)}{\sin x \sin y/[(2\sin^2 x/2 - 1)(2\sin^2 y/2 - 1)] - 1}.$$

**Example 3.25.** For a real variable $x$, convert function $y = |2x^2 - 3| + |4x - 5|$ into a piecewise function.

**Solutions.** The original function can be expressed first in MATLAB, then function `rewrite()` can be used to convert the function into a piecewise one.

```
>> syms x real; y(x)=abs(2*x^2-3)+abs(4*x-5);
   y1=rewrite(y,'piecewise')   % convert to piecewise function
```

the piecewise function obtained is

$$y_1(x) = \begin{cases} 2x^2 + 4x - 8 & \text{if } 5/4 \leqslant x, \\ 2x^2 - 4x + 2 & \text{if } x \leqslant 5/4 \text{ and } 0 \leqslant 2x^2 - 3, \\ -2x^2 - 4x + 8 & \text{if } x \leqslant 5/4 \text{ and } 2x^2 - 3 \leqslant 0. \end{cases}$$

### 3.4.4 Variable substitution of symbolic expressions

In scientific computing, it is usually necessary to convert a function from one independent variable into another, or into another expression. This is referred to as variable substitution. A practical MATLAB function `subs()` is used to perform the conversion, with the syntaxes

$f_1$=subs$(f, x_1, x_1^*)$   % variable substitution, similar to dot operation
$f_1$=subs$(f, \{x_1, x_2, \dots, x_n\}, \{x_1^*, x_2^*, \dots, x_n^*\})$ % several variables

where $f$ is the original expression. In the former syntax, the objective is to substitute $x_1$ in the expression for $x_1^*$, then generate a new expression $f_1$. In the latter one, several variables are substituted simultaneously.

### 3.4.5 Conversions of symbolic expressions

The symbolic expression in MATLAB can be converted to LaTeX typesetting strings, with function `latex()`. The converted strings can be embedded into LaTeX documents, so as to get professional typesetting quality.

**Example 3.26.** Consider the polynomial $P(s)$ in Example 3.21. Substitute $s$ into $s = (z - 1)/(z + 1)$. Convert the result into the LaTeX format.

**Solutions.** The variable substitution like this is referred to as bilinear transformation. The simplest form can be computed directly

```
>> syms s z; P=(s+3)^2*(s^2+3*s+2)*(s^3+12*s^2+48*s+64);
   P1=simplify(subs(P,s,(z-1)/(z+1))), latex(P1)      % substitution
```

The converted string is as follows:

```
\frac{8\, z\, {\left(2\, z + 1\right)}^2\, \left(3\, z + 1\right)
\,{\left(5\, z + 3\right)}^3}{{\left(z + 1\right)}^7}
```

The quality LaTeX display is as follows:

$$P_1(z) = 8\frac{(2z + 1)^2 z(3z + 1)(5z + 3)^3}{(z + 1)^7}.$$

There is no conversion tool for Microsoft Word documents.

Numerator and denominator expressions can be extracted from MATLAB. The function name is `numden()`, with the syntax `[n,d]=numden(f)`, where $f$ is a mathematical expression, $n$ and $d$ are the expressions of the extracted numerator and denominator, respectively.

**Example 3.27.** Consider the substitution result in Example 3.26. Extract the numerator and denominator expressions.

**Solutions.** With the following statements, the numerator and denominator expressions can be directly extracted as $n = 8z(2z + 1)^2(3z + 1)(5z + 3)^3$ and $d = (z + 1)^7$, respectively.

```
>> syms s z;
   P=(s+3)^2*(s^2+3*s+2)*(s^3+12*s^2+48*s+64); % polynomial
   P1=simplify(subs(P,s,(z-1)/(z+1))); [n,d]=numden(P1)
```

## 3.5 Fundamental computations with data

A set of MATLAB functions is provided to process data. Some of the functions are demonstrated with the following examples. Matrices can be selected and other relevant functions can be called so that the results can be observed for better understanding of the functions. In this section, integer rounding functions are discussed first, followed by data sorting, maximum and minimum value extraction, as well as computation of certain statistical quantities. Prime number factorization, permutations, and combinations are addressed.

### 3.5.1 Integer rounding and rationalization of data

A group of MATLAB functions is provided to perform integer rounding in different directions, and the functions are listed in Table 3.1. The meanings and syntaxes are straightforward.

**Table 3.1:** Data rounding and conversions.

| function name | syntaxes | function explanation |
|---|---|---|
| floor() | $n$=floor($x$) | $x$ is rounded in the direction of $-\infty$, with $n$, denoted as $n = \lfloor x \rfloor$ |
| ceil() | $n$=ceil($x$) | rounding $x$ in the direction of $+\infty$, denoted as $n = \lceil x \rceil$ |
| round() | $n$=round($x$) | rounding all the elements in $x$ to the nearest integers, and obtain $n$ |
| fix() | $n$=fix($x$) | rounding $x$ towards zero, obtain $n$ |
| rat() | [$n$,$d$]=rat($x$) | finding the rational expressions of elements in $x$, with numerator $n$ and denominator $d$ extracted |
| rem() | $B$=rem($A$,$C$) | finds the module after division $A$ with respect to $C$ |
| mod() | $B$=mod($A$,$C$) | compute the reminder after division $A$ with respect to $C$ |

**Example 3.28.** For a set of data −0.2765, 0.5772, 1.4597, 2.1091, 1.191, −1.6187, round them using different methods and observe the results, so as to better understand the functions.

**Solutions.** The following statements can be used to round the data:

```
>> A=[-0.2765,0.5772,1.4597,2.1091,1.191,-1.6187];
   v1=floor(A), v2=ceil(A), v3=round(A),
   v4=fix(A) % different rounding functions to understand their results
```

The rounding results are as follows:

$$v_1 = [-1, 0, 1, 2, 1, -2], \quad v_2 = [0, 1, 2, 3, 2, -1],$$
$$v_3 = [0, 1, 1, 2, 1, -2], \quad v_4 = [0, 0, 1, 2, 1, -1].$$

**Example 3.29.** For a $3 \times 3$ Hilbert matrix generated with $A$=hilb(3), find the rationalized matrices.

**Solutions.** With the following statements, the rationalization results can be obtained:

```
>> A=hilb(3); [n,d]=rat(A) % extract numerator and denominator matrix
```

The two integer matrices extracted are

$$n = \begin{bmatrix} 1 & 1 & 1 \\ 1 & 1 & 1 \\ 1 & 1 & 1 \end{bmatrix}, \quad d = \begin{bmatrix} 1 & 2 & 3 \\ 2 & 3 & 4 \\ 3 & 4 & 5 \end{bmatrix}.$$

**Example 3.30.** For the given vector $v = [5.2, 0.6, 7, 0.5, 0.4, 5, 2, 6.2, -0.4, -2]$, find all the integers.

**Solutions.** From function names, it seems that function isinteger() is a suitable one for this example. However, it is not true, since the function can be used to find which one of them is integer-type data, rather than integer. For instance, 7 in the vector is represented as a double precision data, so the result of isinteger() function is 0 rather than 1. Therefore, the function may fail. Remainders should be used instead. With the following statements, it is found that $i = [3, 6, 7, 10]$, indicating the numbers at these locations are integers. From the statements, it can be seen that the integers in the vector are $[7, 5, 2, -2]$, which are the same as those we are expecting.

```
>> v=[5.2 0.6 7 0.5 0.4 5 2 6.2 -0.4 -2];
   i=find(rem(v,1)==0), v(i)   % the simplified statements v(rem(v,1)==0)
```

In fact, the judgement of integers is sometimes not reliable, under the framework of the double precision scheme. For a data of 5.000000000000001, the conditions should be relaxed, such as

```
>> i=find(rem(v,1)<=1e-12), v(i)   % or use smaller number
```

### 3.5.2 Sorting and finding maximum and minimum of vectors

The sorting function sort() can be used to sort the values in the vector in ascending order, with $v$=sort($a$) or [$v$,$k$]=sort($a$). The former returns the sorted vector $v$, while the latter returns also the indexes $k$. In order to sort the data in descending order, the sorting of $-a$ should be used, or an option 'descend' should be given, i. e., with sort($a$,'descend').

**Example 3.31.** If $a$ is a matrix, the function sort() can also be used, each column in the matrix is sorted individually. Let's sorting each column of the $4 \times 4$ magic matrix in ascending orders.

**Solutions.** The matrix can be sorted with the following statements:

```
>> A=magic(4), [a k]=sort(A)
```

and the results are obtained below, together with the related matrices, as

$$
A = \begin{bmatrix} 16 & 2 & 3 & 13 \\ 5 & 11 & 10 & 8 \\ 9 & 7 & 6 & 12 \\ 4 & 14 & 15 & 1 \end{bmatrix}, \quad a = \begin{bmatrix} 4 & 2 & 3 & 1 \\ 5 & 7 & 6 & 8 \\ 9 & 11 & 10 & 12 \\ 16 & 14 & 15 & 13 \end{bmatrix}, \quad k = \begin{bmatrix} 4 & 1 & 1 & 4 \\ 2 & 3 & 3 & 2 \\ 3 & 2 & 2 & 3 \\ 1 & 4 & 4 & 1 \end{bmatrix}.
$$

If $a$ is a matrix, and one wants to sort the rows of the matrix, in ascending order, two methods can be used: the first is to sort the matrix $a^{\mathrm{T}}$, while the other is to use sort$(a,2)$, where 2 represents the sorting with respect to rows. If option 1 is given, the default column-wise sorting is performed. The following statements can be used in the sorting process:

```
>> A, [a k]=sort(A,2)
```

The sorting results are

$$
A = \begin{bmatrix} 16 & 2 & 3 & 13 \\ 5 & 11 & 10 & 8 \\ 9 & 7 & 6 & 12 \\ 4 & 14 & 15 & 1 \end{bmatrix}, \quad a = \begin{bmatrix} 2 & 3 & 13 & 16 \\ 5 & 8 & 10 & 11 \\ 6 & 7 & 9 & 12 \\ 1 & 4 & 14 & 15 \end{bmatrix}, \quad k = \begin{bmatrix} 2 & 3 & 4 & 1 \\ 1 & 4 & 3 & 2 \\ 3 & 2 & 1 & 4 \\ 4 & 1 & 2 & 3 \end{bmatrix}.
$$

If one wants to sort all the elements in the whole matrix, command $A(:)$ can be used to convert the matrix into a column vector. Then, function sort() can be used in the sorting process. This method can also be used in the sorting of a multidimensional array.

```
>> [v,k]=sort(A(:))    % sorting the whole matrix
```

The sorted index vector of $k$ is $k^{\mathrm{T}} = [16, 5, 9, 4, 2, 11, 7, 14, 3, 10, 6, 15, 13, 8, 12, 1]$.

The maximum and minimum values of a matrix can be evaluated with functions max() and min(). The sum and product of matrices can be evaluated with functions sum() and prod(), and the syntaxes are the same for the sorting function.

### 3.5.3 Mean, variance and standard deviation

For a given set of data, statistical analysis is always expected. The definitions of mean, variance, and other statistical quantities are given first, then the computation methods are presented.

**Definition 3.9.** For a given vector $A = [a_1, a_2, \ldots, n]$, the mean $\mu$, variance $\sigma^2$ and standard deviation are defined as:

$$\mu = \frac{1}{n}\sum_{k=1}^{n} a_k, \quad \sigma^2 = \frac{1}{n}\sum_{k=1}^{n}(a_k - \mu)^2, \quad s = \sqrt{\frac{1}{n-1}\sum_{k=1}^{n}(a_k - \mu)^2}. \tag{3.5.1}$$

If $A$ is a vector, functions $\mu$=mean$(A)$, $c$=cov$(A)$, and $s$=std$(A)$ can be used to compute the mean, variance, and standard deviation, respectively, from the vector $A$.

If $A$ is a matrix, the functions act on each column of $A$ independently, the result of mean() function is a vector, containing the means of each column. These functions are identical to those used in the function max(). Each column of matrix $A$ is regarded as a sample of a signal, and function cov() can be used to compute the covariance matrix of the signals.

**Example 3.32.** With function randn(3000,4), four channels of signals can be generated as standard normal distribution signals, each with 3000 samples. Compute the mean of the signals, and compute the covariance matrix of the four signals.

**Solutions.** With powerful computer languages such as MATLAB, the following statements can be written:

```
>> R=randn(3000,4); m=mean(R), C=cov(R)
```

The vector of means is $m = [-0.00388, 0.0163, -0.00714, -0.0108]$, and the covariance matrix is

$$C = \begin{bmatrix} 0.98929 & -0.01541 & -0.012409 & 0.011073 \\ -0.01541 & 0.99076 & -0.003764 & 0.003588 \\ -0.012409 & -0.003764 & 1.0384 & 0.013633 \\ 0.011073 & 0.003588 & 0.013633 & 0.99814 \end{bmatrix}.$$

### 3.5.4 Prime factors and polynomials

The greatest common divisor (GCD) and least common multiple (LCM) can be evaluated from the functions gcd() and lcm(), with the syntaxes $k$=gcd$(n,m)$ and $k$=lcm$(n,m)$. The GCD and LCM are integers for two integers $n$ and $m$, while if $n$ and $m$ are polynomials, the GCD and LCM are also polynomials.

The function $v$=factor$(n)$ can be used to compute prime numbers factorizing $n$. The prime numbers are returned in vector $v$. If $n$ is a polynomial, factorization of the polynomial is computed.

**Definition 3.10.** If the greatest common divisor of polynomials $P(s)$ and $Q(s)$ is independent of $s$, the two polynomials are referred to as coprime.

**Example 3.33.** Compute the greatest common divisor and the least common multiple of two numbers, 1 856 120 and 1 483 720. Compute the factorized form of the latter one.

**Solutions.** Since the numbers are rather large, it is not suitable to process the computation in numerical format, symbolic forms are used instead. The following MATLAB commands can be used to compute directly the results:

```
>> m=sym(1856120); n=sym(1483720);
   gcd(m,n), lcm(m,n), factor(lcm(n,m))
```

The greatest common divisor can be obtained as 1960, and the least common multiple is 1 405 082 840. Prime factorization of the LCM yields $2^3 \cdot 5 \cdot 7^2 \cdot 757 \cdot 947$.

The functions gcd() and lcm() can be used to compute GCDs and LCMs of two integers or polynomials. If more than two arguments are involved, nested calls of the functions should be made with gcd(gcd($m,n$),$k$).

**Example 3.34.** For two given polynomials

$$A(x) = x^4 + 7x^3 + 13x^2 + 19x + 20,$$
$$B(x) = x^7 + 16x^6 + 103x^5 + 346x^4 + 655x^3 + 700x^2 + 393x + 90,$$

judge whether they are coprime or not.

**Solutions.** The solution of the problem can be found employing function gcd(), with the following statements:

```
>> syms x; A=x^4+7*x^3+13*x^2+19*x+20; % the polynomials
   B=x^7+16*x^6+103*x^5+346*x^4+655*x^3+700*x^2+393*x+90;
   d=gcd(A,B)        % compute the greatest common divisor
```

The greatest common divisor can be obtained as $d = x + 5$. Since it is a function of $x$, the two polynomials are not coprime.

**Example 3.35.** List all the prime numbers in the range 1~1 000.

**Solutions.** With the following statements, all the prime numbers can be found, and they are listed in Table 3.2 in a more readable form. In the solution process, isprime($A$) measures all the elements in vector $A$, and sets those primes to the value 1. The statement here is special, and it extracts all the elements in $A$, if isprime($A$) equals one, i. e., all the prime numbers are retained.

```
>> A=1:1000; B=A(isprime(A))   % also a submatrix extraction method
```

A simpler command primes(1000) can be used to extract directly the prime numbers not larger than 1 000.

**Table 3.2:** All prime numbers up to 1 000.

| | | | | | | | | | | | | | | | | | | |
|---|---|---|---|---|---|---|---|---|---|---|---|---|---|---|---|---|---|---|
| 2 | 3 | 5 | 7 | 11 | 13 | 17 | 19 | 23 | 29 | 31 | 37 | 41 | 43 | 47 | 53 | 59 | 61 | 67 |
| 71 | 73 | 79 | 83 | 89 | 97 | 101 | 103 | 107 | 109 | 113 | 127 | 131 | 137 | 139 | 149 | 151 | 157 | 163 |
| 167 | 173 | 179 | 181 | 191 | 193 | 197 | 199 | 211 | 223 | 227 | 229 | 233 | 239 | 241 | 251 | 257 | 263 | 269 |
| 271 | 277 | 281 | 283 | 293 | 307 | 311 | 313 | 317 | 331 | 337 | 347 | 349 | 353 | 359 | 367 | 373 | 379 | 383 |
| 389 | 397 | 401 | 409 | 419 | 421 | 431 | 433 | 439 | 443 | 449 | 457 | 461 | 463 | 467 | 479 | 487 | 491 | 499 |
| 503 | 509 | 521 | 523 | 541 | 547 | 557 | 563 | 569 | 571 | 577 | 587 | 593 | 599 | 601 | 607 | 613 | 617 | 619 |
| 631 | 641 | 643 | 647 | 653 | 659 | 661 | 673 | 677 | 683 | 691 | 701 | 709 | 719 | 727 | 733 | 739 | 743 | 751 |
| 757 | 761 | 769 | 773 | 787 | 797 | 809 | 811 | 821 | 823 | 827 | 829 | 839 | 853 | 857 | 859 | 863 | 877 | 881 |
| 883 | 887 | 907 | 911 | 919 | 929 | 937 | 941 | 947 | 953 | 967 | 971 | 977 | 983 | 991 | 997 | | | |

### 3.5.5 Permutations and combinations

Permutations and combinations are important concepts in combinatorics. They can be computed from their definitions and basic formulas. MATLAB implementation of the methods are presented in this section.

**Definition 3.11.** If we select $k$ different elements from $n$, $k \leqslant n$, and arrange them in a certain order, this kind of arrangement is referred to as a permutation, their number is denoted as $\mathrm{P}_n^k$.

**Definition 3.12.** Selecting $k$ different elements from $n$, $k \leqslant n$, defines a combination, their number is denoted as $\mathrm{C}_n^k$. The number of combinations can be computed from $C$=nchoosek($n,k$) command.

**Theorem 3.2.** *Direct formulas for permutations and combinations are respectively*

$$\mathrm{P}_n^k = \frac{n!}{k!} \quad and \quad \mathrm{C}_n^k = \frac{n!}{k!(n-k)!}. \tag{3.5.2}$$

The number of permutations $\mathrm{P}_n^k$ can be computed directly with the formula in Theorem 3.2. The number of combinations can be evaluated from function nchoosek(). Besides, perms($v$) function can be used to compute all the possible permutations of vector $v$, when $n \leqslant 10$. Function $v$=randperm($n$) can be used to generate a randomly arranged vector $v$ of numbers $1 \sim n$.

**Example 3.36.** There are five persons to arrange to take group photographs. The persons are labeled 1 to 5. List all their possible locations.

**Solutions.** This is a typical permutation problem, since we are not only interested in how many permutations there are, but also what they are. The following commands can be written, and all the permutations are returned in matrix $P$, whose size is 120×5, where $120 = 5!$.

```
>> P=perms(1:5), size(P) % find the number of permutations
```

If the persons are labeled as 'a'~'e', the command $P$=perms('abcde') can be used instead, and again 120 permutations are represented.

**Example 3.37.** Assume that 15 students are arranged for oral examination. Write a random order for the students.

**Solutions.** Function randperm() can be used to complete the task, and a random order of the numbers can be made as

$$v = [9, 14, 2, 12, 4, 6, 11, 5, 13, 10, 3, 8, 1, 7, 15].$$

Each execution result is different, so to show real fairness in the arrangement, apromise with the student can be made, such that the order generated next is the final order.

```
>> randperm(15)   % random order of 1~15
```

## 3.6 Problems

3.1 Input the following complex matrix $A$:

$$A = \begin{bmatrix} 1+4j & 2+3j & 3+2j & 4+1j \\ 4+1j & 3+2j & 2+3j & 1+4j \\ 2+3j & 3+2j & 4+1j & 1+4j \\ 3+2j & 2+3j & 4+1j & 1+4j \end{bmatrix},$$

compute $B = A + A^T$, $C = A + A^H$. Are the obtained matrices $B$ and $C$ symmetric?

3.2 For the given polynomial $f(x) = x^5 + 3x^4 + 4x^3 + 2x^2 + 3x + 6$, let $x = (s-1)/(s+1)$ and convert $f(x)$ into a function of $s$.

3.3 For given functions $f(x) = \dfrac{x \sin x}{\sqrt{x^2 + 2(x+5)}}$ and $g(x) = \tan x$, compute the composite functions $f(g(x))$ and $g(f(x))$.

3.4 For matrix $A$ in Problem 3.1, compute matrices $A^2$, $A.*A$, and $A.^2$.

3.5 For matrix $A$ in Problem 3.1 compute all its fourth roots $A^{1/4}$, and validate the results.

3.6 Consider the complex matrix $A$ in Problem 3.1, find the sum of all the elements whose moduli are larger than 5.

3.7 For the matrices $A$ and $B$ given below, compute respectively $e^{At}$, $\sin At$, $e^{A^2 \cos At^2}$, $A^k$, and $k^A$.

$$A = \begin{bmatrix} -4 & 1 & 0 & -2 \\ -6 & 2 & 0 & -8 \\ -10 & 8 & -3 & -14 \\ -1 & 1 & 0 & -4 \end{bmatrix}, \quad B = \begin{bmatrix} -1-15j & -4j & -14j & 1+4j \\ -1+4j & -1+j & -1+4j & -2j \\ 16j & 4j & -1+15j & -1-4j \\ 4j & 0 & 4j & -1-j \end{bmatrix}.$$

3.8 Since the double precision scheme puts a restriction on a certain numbers of digits, it is not possible to keep high precision results under the framework of this scheme. Compute accurately the with symbolic method the number $C_{50}^{10}$.

3.9 Compute the greatest common divisor of 12! and

$$12\,039\,287\,653\,026\,128\,192\,934.$$

3.10 Factorize the given polynomial and find the simplest result:

$$P(s) = s^{11} + 89s^{10} + 3\,512s^9 + 80\,766s^8 + 1\,196\,322s^7 + 11\,902\,506s^6$$
$$+ 80\,458\,224s^5 + 365\,455\,332s^4 + 1\,079\,056\,341s^3 + 1\,953\,065\,477s^2$$
$$+ 1\,984\,148\,320s + 974\,358\,550.$$

3.11 Compute the greatest common divisor of the two polynomials:

$$P(x) = x^5 + 10x^4 + 34x^3 + 52x^2 + 37x + 10,$$
$$Q(x) = x^5 + 15x^4 + 79x^3 + 177x^2 + 172x + 60.$$

3.12 Find the sum of all the prime numbers less than 10 000. Compute also the prime factorization of the result.

3.13 For a $100 \times 100$ magic matrix, find all the elements larger than 1 000, and set accordingly the elements to 0.

3.14 For a $1\,000 \times 1\,000$ magic matrix, find what the row and column numbers are for the element 34 438.

3.15 Simplify the trigonometric function

$$5\cos^4\alpha\sin\alpha - 10\cos^2\alpha\sin^3\alpha + \sin^5\alpha.$$

3.16 For the following irrational numbers, $\pi$, $\sqrt{2}$, $\sqrt{3}$, e, lg 2, and sin 22°, find their rational approximations and compute the errors.

3.17 How many prime numbers are there in the interval $[1, 1\,000\,000]$? Find the product of all such prime numbers. How many decimal digits are there in this product? Measure the total time needed for all these statements.

3.18 For the given matrix $A$, write the corresponding MATLAB statements, so all the even rows are extracted to form matrix $B$. Use $A$=magic(8) command to generate matrix $A$, and validate the correctness of the above statements.

3.19 Extract the real and imaginary parts of matrix $B$ in Problem 3.7. Also compute its phase matrix.

3.20 Assume that the experimental data is given in Table 3.3. Find the minimum and maximum values and compare the rounding methods. For all the elements in the table, find the mean value and variance.

3.21 Consider the data in Table 3.3. Process the data in the table so that only two decimal digits after the decimal point are kept.

**Table 3.3:** Data for Problem 3.20.

| | | | | | | | | |
|--------|---------|---------|---------|--------|---------|---------|---------|---------|
| 2.6869 | 0.7039 | −1.1656 | 0.7794 | 0.1191 | 1.866 | −1.6775 | 1.969 | −0.6671 |
| −1.9059 | 2.9030 | 2.8706 | 0.9849 | 2.9186 | −1.4595 | 1.5469 | 0.9357 | −0.8233 |
| 0.8607 | −0.0129 | −0.5865 | 0.2939 | 1.8141 | −1.3040 | 1.0276 | 0.2378 | 1.9766 |
| 1.1452 | 2.8639 | 2.8673 | −0.1366 | 1.5918 | 1.9394 | 2.1184 | 0.8581 | −1.7122 |
| 1.5864 | 0.0084 | 0.0875 | 1.9327 | 0.6563 | 1.6381 | −0.0420 | −1.3064 | 0.8062 |

# 4 Flow control structures of MATLAB language

Flow control is a useful way to control the execution of computer statements. As for any programming language, loop, conditional, switch, and trial structures are supported in MATLAB. In this chapter, various flow control possibilities are introduced and demonstrated with examples.

In Section 4.1, two kinds of loop structure are introduced. Iterative methods and vectorized programming are demonstrated, and the eventual target of the latter one is to replace loops, so as to construct high efficiency programs. Conditional structures are presented in Sections 4.2. Piecewise vectorized programming method is also introduced, so as to replace a structure variable to create high efficiency programs. In Sections 4.3 and 4.4, switch and trial structures are also presented.

## 4.1 Loop structures

If one wants to execute a piece of code repeatedly, a loop structure should be used. Two types of loop structure are supported in MATLAB. The first is, for a given $n$, execute a piece of code $n$ times. This kind of loop is the `for` loop, since the keyword `for` should be used. The other loop is the `while` loop. It is a conditional loop, with the keyword `while`. Under a certain condition, a piece of code is executed repeatedly. The two types of loop are studied in this section.

### 4.1.1 The `for` loop structure

The `for` loop is a commonly used loop structure, with typical syntax of

```
for i = v, loop body, end
```

In the standard `for` loop structure, $v$ is a vector. Loop variable $i$ takes an element from vector $v$, and executes the "loop body", then returns to `for` command, takes another element from vector $v$, assigns it to $i$, and executes "loop body" again. The process is executed again and again, until all the elements in vector $v$ are selected. The loop structure is then completed.

If $v$ is a matrix, then each time $i$ takes a column vector, until all the columns in the matrix are executed.

**Example 4.1.** Consider first a simple example. Compute the sum $S = \sum_{i=1}^{100} i$ with a loop structure.

**Solutions.** The `for` loop can be used to compute the sum. Assign $v$ as a row vector $[1, 2, \ldots, 100]$, and also consider a cumulative variable $s$, with the initial value of 0. In

https://doi.org/10.1515/9783110666953-004

the for loop, each element in vector $v$ is selected, and added to the variable $s$. With the code below, the final result of $s = 5\,050$ can be found.

```
>> s=0; for i=1:100, s=s+i; end, s  % simple loop structure
```

In fact, the sum can easily be obtained with sum(1:100). The same result can be obtained. The sum() function can be used to act on the whole vector, so that the program is simplified.

It can be seen that the for loop structure here is more flexible than that in C language. In the next example, the for loop is used in an iterative process.

**Example 4.2.** Consider a sequence, whose the first term is $a_1 = 3$, and the second term can be computed from the first term, $a_2 = \sqrt{1 + a_1}$. The subsequent terms can be expressed by an iterative formula $a_{k+1} = \sqrt{1 + a_k}$, $k = 1, 2, \ldots, m$. Compute $a_{32}$ of the sequence.

**Solutions.** This kind of process is referred to as an "iterative" process. For convenient programming, a vector can be generated to store $a_k$, such that $a(1) = 3$. Executing the loop of $k$ for 31 times, the term $a(32)$ can be computed. The following loop structure can be used to compute the sequence:

```
>> a(1)=3; format long                    % set display format
   for k=1:31, a(k+1)=sqrt(1+a(k)); end, a % iteration process
```

It can be seen that the 31st and 32nd terms are both 1.618033988749895, i. e., they are identical. Further iteration may not improve the result under double precision framework. The iteration process is regarded as convergent. The limit value is also known as the golden ratio[12].

The sequence here can be understood in the view of mathematics, and the first few terms can be expressed as

$$3, \quad \sqrt{1 + 3}, \quad \sqrt{1 + \sqrt{1 + 3}}, \quad \sqrt{1 + \sqrt{1 + \sqrt{1 + 3}}}, \ldots$$

Assuming that the sequence converges to $x$, it can be seen from the iteration formula that $x = \sqrt{1 + x}$. Solving the equation, one may find $x = (1 + \sqrt{5})/2 \approx 1.618033988749895$.

**Example 4.3.** For a $9 \times 9$ magic matrix, compute the sum of each row of the matrix.

**Solutions.** The following statements can be used to generate a magic matrix.

```
>> A=magic(9)
```

It can be seen that the magic matrix is

$$A = \begin{bmatrix} 47 & 58 & 69 & 80 & 1 & 12 & 23 & 34 & 45 \\ 57 & 68 & 79 & 9 & 11 & 22 & 33 & 44 & 46 \\ 67 & 78 & 8 & 10 & 21 & 32 & 43 & 54 & 56 \\ 77 & 7 & 18 & 20 & 31 & 42 & 53 & 55 & 66 \\ 6 & 17 & 19 & 30 & 41 & 52 & 63 & 65 & 76 \\ 16 & 27 & 29 & 40 & 51 & 62 & 64 & 75 & 5 \\ 26 & 28 & 39 & 50 & 61 & 72 & 74 & 4 & 15 \\ 36 & 38 & 49 & 60 & 71 & 73 & 3 & 14 & 25 \\ 37 & 48 & 59 & 70 & 81 & 2 & 13 & 24 & 35 \end{bmatrix}.$$

If command sum($A$) is used, all the column elements in $A$ can be added to form a vector. However, what we are expecting is the sums of the rows, rather than of columns. Therefore, the transpose of $A$ should be taken first, then function sum() can be called. For this example, a loop structure can be used to add elements of each row.

A zero $9 \times 1$ vector $s$ can be generated first. In the loop structure, $i = A$, since $A$ is a matrix, the vector $i$ takes a column vector from $A$ each time, then the sum can be computed. It can be seen that the following statements can be used such that the sums of all the elements in each row can be obtained, and they are all 369.

```
>> s=0; for i=A, s=s+i; end, s, sum(A.').' % the two results are the same
```

**Example 4.4.** For the Fibonacci sequence, the general term is $a_k = a_{k-1} + a_{k-2}$, $k = 3, 4, \ldots$, where the initial values are $a_1 = a_2 = 1$. Compute the first 100 terms of the Fibonacci sequence.

**Solutions.** It can be seen from the Fibonacci sequence formula that, if a vector is used to describe the sequence, the first two terms are $a(1) = a(2) = 1$, and from the third term onwards, an iterative formula can be introduced, so the following statements can be used to compute the Fibonacci sequence:

```
>> a=[1 1]; for k=3:100, a(k)=a(k-1)+a(k-2); end, a(end)
```

Unfortunately, it can be seen from the results that the double precision scheme can only research 15 decimal digits. Therefore the result obtained is not accurate. The symbolic data structure can be used instead. To use symbolic data type, the only change is the initial term, which can be converted to a symbolic one. It can be seen that the result is then $a_{100} = 354\,224\,848\,179\,261\,915\,075$.

```
>> a=sym([1 1]); % change the initial values to symbolic ones
   for k=3:100, a(k)=a(k-1)+a(k-2); end, a(end)
```

In for loops, a predefined condition can be given, then the loop variable can be assigned. The for loop can be regarded as an unconditional loop structure.

### 4.1.2 The `while` loop structure

The `while` loop is another type of loop structure. Unlike the unconditional `for` loop, the `while` loop allows conditions. If a condition is not true, the loop is terminated. The typical form of the `while` loop is

`while (condition), loop body, end`

In the `while` loops, the condition expression in the `while` statement is assessed first, if it is true, the loop can be executed once, then we go back to the `while` statement. The condition expression is evaluated again. If it is true, the loop is executed again, and if the condition is not true, the loop structure is terminated. If the condition is not a logic variable, then, if the expression is not zero, it can be regarded as true, otherwise it is false, to terminate the loop.

The loop structures are led by `for` or `while` statements, and they must be ended with an `end` statement. The statements in-between are referred to as "loop body". The forms and applications of the two statements are different. Examples are given below to demonstrate the differences.

**Example 4.5.** Compute $S = \sum_{i=1}^{100} i$ again with the `while` loop.

**Solutions.** Similar to the `for` loop, a cumulative variable $s$ with zero initial value is created. Meanwhile, in the `for` loop, $i$ is a dependent variable, which takes values from vector $v$. In the `while` loop, $i$ is an independent variable, with initial value of zero. The condition in the `while` loop is to test whether $i \leqslant 100$ is satisfied. If so, $i$ is added to $s$, and also it is incremented by 1, and we go back to the `while` statement to test again whether $i \leqslant 100$ is satisfied or not. If the condition is satisfied, the terms are added on and on, until the condition is not true, meaning all the 100 terms are already added up. The loop can then be completed. The sum found is $s = 5\,050$, which is identical to the previous result.

```
>> s=0; i=1;
   while (i<=100), s=s+i; i=i+1; end, s
```

Comparing the `for` and `while` loops, it can be seen that, for this particular example, the `for` loop is easier.

**Example 4.6.** Find the smallest $m$ such that $S = \sum_{i=1}^{m} i > 10\,000$.

**Solutions.** For this example, it can be seen that the `for` loop structure is not suitable, since before the accumulation process, we do not know how many terms are to be added. Therefore, unconditional loop fails in this type of problem. The `while` loop can be used instead for this problem. The independent variable $m$ can be added to the cumulative variable $s$, and in each step in the loop, the value of $s$ is computed and checked whether $s \leqslant 10\,000$ or not. If it is true, the terms can be added to $s$, and if it is false, the loop is completed, to indicate that the value of $m$ is found. With the above ideas in mind, the following statements can be written, and the result is $s = 10\,011$, $m = 141$. The result can be validated with $\mathrm{sum}(1:m)$ command.

```
>> s=0; m=0;
   while (s<=10000), m=m+1; s=s+m; end, s, m % sum is larger than 10 000
```

### 4.1.3 Loop implementation of iterations

Iteration is a very important approach in numerical computations. In real applications, the same piece of code is repeatedly executed. The error can be reduced in the loop structure, so as to approach the solution of the problem. Loop structure is suitable for implementing iterative processes. The whole iterative process is represented by a loop framework. If a certain condition is satisfied, the loop can be terminated. The iteration process is demonstrated through examples.

**Example 4.7.** It is known that $\arctan x = x - x^3/3 + x^5/5 - x^7/7 + \cdots$. Letting $x = 1$, the following formula can be obtained:

$$\pi \approx 4\left(1 - \frac{1}{3} + \frac{1}{5} - \frac{1}{7} + \frac{1}{9} - \frac{1}{11} + \cdots\right).$$

Use an iterative process to compute $\pi$, with an accuracy of $10^{-6}$.

**Solutions.** To solve this kind of problem, the general term in the parentheses can be formulated. The $k$th term can be expressed as $s_k = (-1)^{k+1}/(2k-1)$, $k = 1, 2, \ldots$. Since it is not known how many terms are to be accumulated, the $\mathrm{for}$ loop structure cannot be used. The $\mathrm{while}$ structure can be used to deal with the problem, and the absolute value of the cumulative term $s_1$ can be used as the termination condition. If its value is less than or equal to $10^{-6}$, the loop can be terminated. The numerical approximation of $\pi$ can be obtained as 4 times the cumulative sum, where $S = 3.141594653585692$. The cumulative step is $k = 500\,002$, and the time needed is 0.04 seconds.

```
>> s=0; k=1; s1=1; tic
   while (abs(s1)>1e-6),
       s1=(-1)^(k+1)/(2*k-1); k=k+1; s=s+s1;
   end, S=4*s, k, toc
```

It can be seen that the accuracy is higher than what is expected, since the first 11 digits are valid, meaning that the termination condition is too strict. In fact, if one wants to have an accuracy of $10^{-6}$, the termination condition can be changed to $4|s - s_0| \leqslant 10^{-6}$, where $s$ is the current result, and $s_0$ was the result in the last iterative step. To enable the while loop structure, set the initial value of $s_0 \neq 0$. For instance, let $s_0 = 1$. The whole loop structure can be modified as follows, and the approximation of $\pi \approx$ 3.1415931 can be obtained. The number of steps is reduced to 125 002, with time of 0.016 seconds.

```
>> s=0; s0=1; k=1; tic
   while (abs((s-s0)*4)>1e-6), s0=s;
       s1=(-1)^(k+1)/(2*k-1); k=k+1; s=s+s1;
   end, S=4*s, k, toc
```

**Example 4.8.** The efficiency of different algorithms used to approximate $\pi$ is usually low. Another algorithm to approximate $\pi$ is proposed as follows:

$$\frac{2}{\pi} \approx \frac{\sqrt{2}}{2} \cdot \frac{\sqrt{2 + \sqrt{2}}}{2} \cdot \frac{\sqrt{2 + \sqrt{2 + \sqrt{2}}}}{2} \cdots$$

When the terminating condition is selected as $|\delta - 1| < \epsilon$, where $\delta$ is the factor, and a more accurate condition $\epsilon = 10^{-15}$ is selected, which is much stricter than that used in Example 4.7, find the approximate value of $\pi$, and comment on the result.

**Solutions.** The problem now can only be solved with the while loop structure, rather than the for loop structure. Setting the result variable as $P = 1$, the general term can be represented by the recursive formula $\sqrt{2 + d_0} \to d_0$. Therefore, in each step of the loop, the multiplier can be updated to $d = d_0/2$. The distance between the multiplier and 1 can be used in the termination condition, e. g., $|d - 1| \leqslant \epsilon$. If it is satisfied, the loop can be terminated, and the product can be regarded as the approximation of $\pi$. It can be seen that only $k = 28$ iteration steps are needed, and one finds $\pi \approx 3.141592653589794$. The total time needed is only 0.0072 seconds. The efficiency and accuracy are much higher than those of the algorithm implemented in Example 4.7.

```
>> tic, P=1; d=2; d0=sqrt(2); k=1;
   while(abs(d-1)>eps),
       d=d0/2; P=P*d; d0=sqrt(2+d0); k=k+1;
   end, toc, 2/P, k
```

**Example 4.9.** The sine function of a matrix can be expressed as the following power series expansion:

$$\sin \boldsymbol{A} = \sum_{k=0}^{\infty} (-1)^k \frac{\boldsymbol{A}^{2k+1}}{(2k + 1)!} = \boldsymbol{A} - \frac{1}{3!}\boldsymbol{A}^3 + \frac{1}{5!}\boldsymbol{A}^5 + \cdots \tag{4.1.1}$$

For the matrix $\boldsymbol{A}$ in Example 3.18, compute $\sin \boldsymbol{A}$ from the above formula with an iterative process.

**Solutions.** Cumulative method can also be used in this example to compute recursively the matrix function. The most important step in the iterative process is to write the general term. It is fine to use $(-1)^k A^{2k+1}/(2k+1)!$ as the general term, however, the power of the matrix should be computed repeatedly. A better way is to define the increment of the next term with respect to the current term, and use a loop structure to construct the general term. For this example, the $k$th term can be expressed as

$$F_k = (-1)^k \frac{A^{2k+1}}{(2k+1)!}, \quad k = 0, 1, 2, \ldots \tag{4.1.2}$$

Therefore the ratio of the next term with respect to the current can be formulated as follows (for simplicity, matrix division is denoted as scalar division):

$$\frac{F_{k+1}}{F_k} = \frac{(-1)^{k+1}A^{2(k+1)+1}/(2(k+1)+1)!}{(-1)^k A^{2k+1}/(2k+1)!} = -\frac{A^2}{(2k+3)(2k+2)}. \tag{4.1.3}$$

An iterative method can be used to implement the sum of the power series of the matrix sine function. If the error is small enough, the accumulation process can be terminated. Here, the termination condition can be selected as $\|E + F - E\|_1 > 0$. The physical interpretation is that the contribution of $F$, when added to the result $E$, can be totally neglected. It is not recommended to simplify it mathematically to $\|F\|_1 > 0$, since it is a notation under the double precision framework. The mathematical notation $\| \cdot \|$ means the norm of the matrix, and it can be computed directly with `norm()` function.

```
>> A=[-1,-1,1,0; -1,-1,-1,0; 1,2,-1,1; 0,-1,0,-2];
   F=A; E=A; k=0;
   while norm(E+F-E,1)>0,
      F=-A^2*F/(2*k+3)/(2*k+2); E=E+F; k=k+1;
   end, E, k
```

The result obtained is as follows, and the number of iterations is $k = 12$:

$$E = \begin{bmatrix} 0.0517403714464 & 0.352909050386 & 0.961037798272 & 0.420735492404 \\ -0.96103779827 & -1.26220647721 & -0.96103779827 & -0.420735492404 \\ -0.35290905039 & 0.187393255482 & -1.26220647721 & 0.119566813464 \\ 0.420735492404 & -0.11956681346 & 0.420735492404 & -0.96103779827 \end{bmatrix}.$$

In fact, the sine of matrix $A$ can be obtained also with `funm(A,@sin)`. The result is identical to that obtained in the example.

### 4.1.4 Assistant statements of loop structures

Loops can be nested in MATLAB programming. The `for` loops can be nested within `while` loops, and vise versa. To finish a loop, there should be a matching `end` statement.

The following assistant statements can be used inside a loop structure:

(1) `break` statement. The `break` statement can be placed anywhere in a loop structure. It can be used to terminate the current-level of a loop structure.
(2) `continue` statement. If the `continue` statement is encountered in the loop structure, the subsequent statements in the loop body are abandoned and one returns to the starting point of the loop structure for the next loop execution.
(3) `return` statement. This statement is not a dedicated one for a loop structure. It can be used anywhere in a MATLAB function. If this statement is encountered, the remaining statements in the whole function are abandoned, and one returns to the calling function.

The assistant statements should be accompanied by other statements, and this will be demonstrated later in examples.

### 4.1.5 Vectorized implementation of loops

In MATLAB programming, the execution of loops is usually slow. Therefore, in actual programming, the matrix or vector can be treated as a whole object, to avoid the usage of loops. This kind of treatment is referred to as vectorized programming. The efficiency of code can be increased.

Vectorized programming in MATLAB is an attractive topic, since vectorized programs usually are considered as aesthetically attractive to experienced MATLAB programmers, while the programs with unnecessary loops are usually considered as low-standard ones. The following examples are used to demonstrate the differences in loops and vectorized programming.

**Example 4.10.** Assuming that there is a set of circles with radii

$$r = [1.0, 1.2, 0.9, 0.7, 0.85, 0.9, 1.12, 0.56, 0.98],$$

compute the areas of the circles.

**Solutions.** For the beginners of MATLAB with experience in C or other programming languages, the following statements can be used:

```
>> r=[1.0,1.2,0.9,0.7,0.85,0.9,1.12,0.56,0.98];
   for i=1:length(r), S(i)=pi*r(i)^2; end, S
```

These commands can indeed be used to compute correctly the areas of the circles, however, they are not decent MATLAB programming. Vectorized programming is recommended for solving this kind of problems. Without using loop structures, the following statement can be used to yield the same results. The statement looks much more beautiful.

```
>> S=pi*r.^2
```

**Example 4.11.** Compute the sum of the series

$$S = \sum_{i=1}^{10\,000\,000} \left( \frac{1}{2^i} + \frac{1}{3^i} \right).$$

**Solutions.** For this example, one may refer to the loops in Example 4.1, to compute the sum directly. The sum obtained is 1.5, and the total time needed is about 1.96 seconds.

```
>> N=10000000;
   tic, s=0; for i=1:N, s=s+1/2^i+1/3^i; end; toc % ordinary loop
```

Now, constructing a row vector $i$, the mathematical expression $1/2^i$ can be implemented with a dot operation as `1./2.^i`, which is still a vector. Similarly, the mathematical expression $1/3^i$ can be evaluated from `1./3.^i`. Adding all the terms in the vector `1./2.^i+1./3.^i`, with `sum()` function, the problem can be solved and a loop structure can be avoided. This is the vectorized programming. The sum is also 1.5, and the total time needed is reduced to 0.64 seconds.

```
>> tic, i=1:N; s=sum(1./2.^i+1./3.^i); toc % vectorized programming
```

In this example, the efficiency of the programming is significantly increased by introducing the vectorized programming technique. In fact, the efficiency of loops in new versions of MATLAB is improved. Compared to the effectiveness in the old versions, the difference is much more significant.

**Example 4.12.** Hilbert matrices are a class of special matrices, whose general term is $h_{ij} = 1/(i + j - 1)$, where $i$ and $j$ are the row and column numbers. An $n \times n$ square Hilbert matrix can be generated with MATLAB function `hilb()`. This function cannot be used to generate rectangular matrices. Generate now a $50\,000 \times 50$ rectangular Hilbert matrix.

**Solutions.** It is obvious that the following double loop structure can be used in generating the expected Hilbert matrix, and the time required is 33.92 seconds.

```
>> tic, for i=1:50000, for j=1:50, H(i,j)=1/(i+j-1); end, end, toc
```

If a $50\,000 \times 500$ matrix is expected, the above method cannot be used. If the order of the loops is swapped, the large loop can be used in the inner loop. The total time is reduced to 0.49 seconds. It can be seen that for the same problem, the larger loop can be used in the inner one, while the smaller is moved to be the outer one. In this case, the efficiency can be increased.

```
>> tic, for j=1:500, for i=1:50000, H1(i,j)=1/(i+j-1); end, end, toc
```

If the inner loop is replaced by a vectorized statement, the time is further reduced to 0.33 seconds.

```
>> tic, for j=1:500, i=1:50000; H2(i,j)=1./(i+j-1); end, toc
```

If both loops are replaced by vectorized forms, the total time is reduced to 0.11 seconds.

```
>> tic, [i,j]=meshgrid(1:50000,1:500); H3=1./(i+j-1); toc
```

**Example 4.13.** Function `meshgrid()` was used in the previous example. Display and observe the results of such a function.

**Solutions.** Function `meshgrid()` can be used to generate two-dimensional mesh grid data, which can be regarded as two-dimensional coordinates. Assuming that the horizontal axis is marked by $v_1 = [1, 2, 4, 3, 5, 7, 9]$, while the vertical axis is marked by $v_2 = [-1, 0, 2]$, the following statements can be used to call `meshgrid()` function, and two matrices can be generated:

```
>> v1=[1 2 3 4 5 7 9]; v2=[-1 0 2]; % horizontal and vertical axes
   [x,y]=meshgrid(v1,v2)            % generate mesh grid matrices
```

The two mesh grid matrices can be generated as follows, from which the facilities in `meshgrid()` function can be understood:

$$x = \begin{bmatrix} 1 & 2 & 4 & 3 & 5 & 7 & 9 \\ 1 & 2 & 4 & 3 & 5 & 7 & 9 \\ 1 & 2 & 4 & 3 & 5 & 7 & 9 \end{bmatrix}, \quad y = \begin{bmatrix} -1 & -1 & -1 & -1 & -1 & -1 & -1 \\ 0 & 0 & 0 & 0 & 0 & 0 & 0 \\ 2 & 2 & 2 & 2 & 2 & 2 & 2 \end{bmatrix}.$$

**Example 4.14.** It was indicated in Example 2.2 that the result of `vpa()` function, when displayed in the form of a command line, may display at most 25 000 characters, the remaining ones cannot be displayed at all. How can we display the first 1 000 000 digits of the irrational value $\pi$?

**Solutions.** A piecewise display format can be employed. Function `vpa()` can be used to extract as many characters as possible, if it is returned to a variable, rather than displaying in a command window. In that case, the string for all the digits can be extracted, and all the digits can be displayed in different lines, with 10 000 characters in each line. It is obvious that a loop structure can be used to extract the digits of $\pi$ into strings, and the results can be displayed with the following statements:

```
>> P=vpa(pi,1000000); str=char(P); n=10000;
   for i=1:n:length(str)
      fprintf('%s\n',str(i:min(i+n-1,length(str))));
   end
```

The expected digits of $\pi$ can be displayed in 101 lines, and in the last line, there is only one digit, since the total number of digits is 1 000 001, because the decimal point holds one place.

If one is not familiar with `fprintf()` function, the whole command can be replaced by `disp(str(i:min(i+n-1,length(str))))`.

## 4.2 Conditional structures

Conditional structures are supported in ordinary programming languages. Normally the conditions are provided to determine which branch of the program to execute. Under different conditions, different tasks are executed in the problem. The basic conditional structure is `if ... end`, and sometimes, the keywords `else` and `elseif` can be used to extend the conditional structures. In this section, various conditional structures are discussed, and examples are used to illustrate them.

### 4.2.1 Simple conditional structures

The simplest form of conditional structure is

`if (condition), statement group, end`

where "condition" is an expression. The physical interpretation of the structure is as follows: if "condition" is true, the commands in the "statement group" are executed. After that, the `end` keyword is executed, which means the end of the structure. If the condition is not true, then the "statement group" can be bypassed.

Another simple conditional structure is expressed as

`if (condition), statement group 1, else, statement group 2, end`

Similar to the simple case, if the condition is true, the "statement group 1" is executed, otherwise, the "statement group 2" is executed. After the execution, the whole structure is completed. Remember to use the matching keyword `end` in the structure.

### 4.2.2 General form of conditional structures

Apart from the simple conditional structures, a more general conditional structure in MATLAB is as follows:

```
if (condition 1) % if condition 1 is satisfied, execute statement group 1
        statement group 1 % lower-level if structure can be embedded
    elseif (condition 2) % if condition 2 is satisfied instead, then group 2
        statement group 2
        ⋮ % many such conditions can be used
    else % if none of the conditions are satisfied, then execute the following
        statement group n + 1
end
```

**Example 4.15.** Structures of the `for` loop and `if` statement can be used together to solve again the problem in Example 4.6.

**Solutions.** In Example 4.6, it was stated that by using the `for` loop alone, the problem cannot be solved. When the `for` loop is used with an `if` structure, the problem can be solved. More specifically, the `for` loop can be run in a large interval of $i$. The value of $i$ can be accumulated to the variable $s$. Then, the sum is checked whether it is larger than 10 000. If not, we continue the loop process, otherwise, statement `break` is used to exit from the loop structure. The needed sum is then obtained. For following MATLAB statements can be used:

```
>> s=0; for i=1:10000, s=s+i; if s>10000, break; end, end
```

It can be seen that the structure is not as simple as the `while` structure, presented earlier in Example 4.6.

**Example 4.16.** Bisection method is an easy-to-understand algorithm for solving and algebraic equation $f(x) = 0$. Assume that in an interval $(a, b)$, one has $f(a)f(b) < 0$, and the function is continuous, then, there exists at least one solution in the interval. Take the middle point of the interval $x_1 = (a + b)/2$. The interval of a solution can be determined from the signs of the function values at $f(x_1), f(a),$ and $f(b)$. In this way, the length of the interval can be reduced by half. Repeating the process, until the length of the interval is smaller than the preselected small value $\epsilon$, the middle point can be regarded as the solution of the equation. Selecting $\epsilon = 10^{-10}$, solve approximately the algebraic equation $f(x) = x^2 \sin(0.1x + 2) - 3 = 0$ in the interval $(-4, 5)$.

**Solutions.** The algebraic equation can be described by an anonymous function (details of anonymous functions will be presented later).

```
>> f=@(x)x.^2.*sin(0.1*x)-3;
```

A loop structure can be used in solving the algebraic equation, with the bisection algorithm. If the length of the interval is greater than $\epsilon$, the loop is executed automatically, until the length of the interval is smaller than $\epsilon$, and the loop can be completed successfully. In the loop, the middle point $x = (a + b)/2$ is selected. If the signs of $f(x)$ and $f(a)$ are different, it means that there is at least one solution in the interval $(a, x)$. The terminal $b$ can be set to $x$, otherwise, $a$ can be set to $x$. In this way, the length of the interval is reduced by half, and the loop is continued. After a few interval-halving iterations, the remaining length gets so small that the condition $|a - b| > \epsilon$ is no longer satisfied. The solution of the equation can be found.

Bisection algorithm can be implemented with the following statements, and it can be used to solve the original equation:

```
>> a=-4; b=5; err=1e-10; k=0; tic
   while abs(b-a)>err, x=(a+b)/2; k=k+1;
       if f(x)*f(a)<0, b=x; else, a=x; end
   end, x, f(x), k, toc
```

The solution obtained is $x = 3.124182730367465$, the number of loop iterations is $k = 37$, and total time needed is 0.0141 seconds. If the result is substituted back to the equation, the error is $f(x) = -8.4361 \times 10^{-11}$. This means that the solution approximately satisfies the original equation. The solution scheme here is feasible. For this example, it is not suitable to use an even smaller error tolerance `err`.

**Example 4.17.** Consider the coin-tossing problem, and simulate it on a computer. If a coin is tossed 100 000 times, how many of them result face-up?

**Solutions.** Pseudorandom numbers can be used to carry out the experiment. A set of pseudorandom numbers can be generated with function `rand()`, and it is assumed that the numbers have the uniform distribution in the interval $(0, 1)$. Denote the random numbers as "face-up", when the value is larger than 0.5, otherwise call them "face-down". Then, the coin-tossing experiment can be completed with the following statements, and it can be found that the number of "face-up" cases is 50 132, around half of the tossings.

```
>> r=rand(1,100000); k=0;
   for i=1:length(r), if r(i)>=0.5, k=k+1; end, end, k
```

Since the experiment here is a random one, the results in each run of the code are not exactly the same, however, they are similar. In fact, the conditionals and loop structures can be avoided, if the following statement is used, where function $\mathrm{nnz}(\boldsymbol{A})$ is employed to count the number of nonzero elements in matrix $\boldsymbol{A}$:

```
>> k1=nnz(r>=0.5)
```

**Example 4.18.** Factorial tree is a useful example in showing the conditional structures in MATLAB. For an arbitrary initial point $(x_0, y_0)$ in a two-dimensional plane, assume that a pseudorandom number $y_i$ is generated, which is uniformly distributed in the interval $[0, 1]$. A new point $(x_1, y_1)$ can be computed from the following formula[22]:

$$(x_1, y_1) \Leftarrow \begin{cases} x_1 = 0, \quad y_1 = y_0/2, & \text{if } y_i < 0.05, \\ x_1 = 0.42(x_0 - y_0), \quad y_1 = 0.2 + 0.42(x_0 + y_0), & \text{if } 0.05 \leqslant y_i < 0.45, \\ x_1 = 0.42(x_0 + y_0), \quad y_1 = 0.2 - 0.42(x_0 - y_0), & \text{if } 0.45 \leqslant y_i < 0.85, \\ x_1 = 0.1x_0, \quad y_1 = 0.2 + 0.1y_0, & \text{otherwise,} \end{cases}$$

and the new point can be used as the initial point to generate the next point. In this way, generate a set of 10 000 points.

**Solutions.** A loop structure can be used to generate the 10 000 points from the initial one $(x_0, y_0)$. Assuming that a set of pseudorandom numbers is generated, and each new point can be computed with conditional statements, the set of points can be generated with the following MATLAB statements:

```
>> v=rand(10000,1); N=length(v); x=0; y=0;
   for k=2:N, gam=v(k);
       if gam<0.05, x(k)=0; y(k)=0.5*y(k-1);
       elseif gam<0.45,
           x(k)=0.42*(x(k-1)-y(k-1)); y(k)=0.2+0.42*(x(k-1)+y(k-1));
       elseif gam<0.85,
           x(k)=0.42*(x(k-1)+y(k-1)); y(k)=0.2-0.42*(x(k-1)-y(k-1));
       else, x(k)=0.1*x(k-1); y(k)=0.1*y(k-1)+0.2;
   end, end
```

### 4.2.3 Vectorized expressions of piecewise functions

Consider the following typical piecewise function:

$$f(x) = \begin{cases} f_1(x), & \text{if } x \geqslant 2, \\ f_2(x), & \text{otherwise.} \end{cases}$$

If the independent variable $x$ is a vector $x$, composed of samples, how can we compute vector $y$ for the given function? For the beginners of MATLAB, having experience in C or other languages, it is quite natural to use a loop for the problem, and for each point $x_i$ in the loop, a conditional structure can be used to compute the value of the function.

If fact, if the reader knows the meaning of the expression $x>=2$ in MATLAB, a more concise vectorized form can be used in the evaluation of piecewise functions. The statement $x>=2$ returns a vector, whose length is the same as of vector $x$, containing 0's and 1's only. For the points where $x \geqslant 2$ is satisfied, the values are 1, otherwise the values are 0. Besides, the word "otherwise" here means logically the condition $x < 2$. The two conditions are referred to as mutually exclusive, since when one of them is true, the other must be false.

With the mutually exclusive logic relationship, piecewise functions can be evaluated in the vectorized form of

$$y=f_1(\mathbf{x}).*(\mathbf{x}>=2)+f_2(\mathbf{x}).*(\mathbf{x}<2)$$

In this case, the loops and conditional structures can be avoided, since the dot operations of the vectors can be used to compute directly the piecewise functions. This method can also be used in dealing with numerical evaluation problems of multidimensional functions. However, one must be sure that the logic conditions are mutually exclusive, otherwise, the results obtained may be wrong.

Besides, for symbolic computation problems of piecewise functions, a MATLAB function piecewise() can be used, with the syntax

$f$=piecewise(condition$_1$,fun$_1$,condition$_2$,fun$_2$,...,condition$_m$,fun$_m$)

where condition$_i$ is the symbolic condition, while fun$_i$ is the function expression. When using the function, make sure that the conditions and function expressions appear in pairs, otherwise error messages may be given.

**Example 4.19.** Express the saturation nonlinear function in MATLAB:

$$y = \begin{cases} 1.1\,\mathrm{sign}(x), & |x| > 1.1, \\ x, & |x| \leqslant 1.1. \end{cases}$$

**Solutions.** Under the symbolic framework, the following statements can be used to express the piecewise function:

```
>> syms x
   f(x)=piecewise(abs(x)>1.1,1.1*sign(x),abs(x)<=1.1,x);
```

If $|x| \leqslant 1.1$, the mathematical expression can be written as $-1.1 \leqslant x \leqslant 1.1$, which can be further expressed as $x \geqslant -1.1$ and $x \leqslant 1.1$. The symbolic expression can be described as x>=-1.1 & x<=1.1.

If a set of samples is stored in a vector $\mathbf{x}$, then a dot operation can be used to compute the vector $\mathbf{y}$, with the following statements:

```
>> x=-2:0.01:2; y=1.1*sign(x).*(abs(x)>1.1)+x.*(abs(x)<=1.1);
```

**Example 4.20.** An example of two-dimensional piecewise function is described as[2]

$$p(x_1,x_2) = \begin{cases} 0.5457\exp(-0.75x_2^2 - 3.75x_1^2 - 1.5x_1), & x_1 + x_2 > 1, \\ 0.7575\exp(-x_2^2 - 6x_1^2), & -1 < x_1 + x_2 \leqslant 1, \\ 0.5457\exp(-0.75x_2^2 - 3.75x_1^2 + 1.5x_1), & x_1 + x_2 \leqslant -1. \end{cases}$$

Represent such an equation in symbolic and numerical statements.

**Solutions.** With the `piecewise()` function, the following statements can be used to describe symbolically the given piecewise expression as

```
>> syms x1 x2;
   p(x1,x2)=piecewise(x1+x2>1,...
        0.5457*exp(-0.75*x2^2-3.75*x1^2-1.5*x1),...
        -1<x1+x2 & x1+x2<=1,0.7575*exp(-x2^2-6*x1^2),...
        x1+x2<=-1,0.5457*exp(-0.75*x2^2-3.75*x1^2+1.5*x1))
```

For a set of mesh grid data matrices $x_1$ and $x_2$, the following statements can be used to compute using a dot operation:

```
>> [x1,x2]=meshgrid(-2:0.1:2,-1.5:0.1:1.5);
   y=0.5457*exp(-0.75*x2.^2-3.75*x1.^2-1.5*x1).*(x1+x2>1),...
    0.7575*exp(-x2.^2-6*x1.^2).*(-1<x1+x2 & x1+x2<=1),...
    0.5457*exp(-0.75*x2.^2-3.75*x1.^2+1.5*x1).*(x1+x2<=-1));
```

**Example 4.21.** Is it possible to solve the problem in Example 4.18 with vectorized statements for the loops and conditional structures?

**Solutions.** Since in the mathematical formula, each new point depends on its previous position, it is not possible to use the vectorized form in the loop programming. The internal conditional structure can be implemented in the vectorized form, however, since each point is a scalar, it is not necessary to do so in the vectorized format.

## 4.3 Switch structures

The switch structure is similar to the case of multiport switches in real circuits. If the switch is dialed to a certain position, then a loop is connected, while the other ones are switched off. The switch structure in programming behaves virtually the same. Switch structures can also be implemented with `if`, `elseif`, `end` structure. For instance, the following nested form is given:

```
if key==expression 1,  statement group 1;
elesif key==expression 2,  statement group 2;
    ⋮
end
```

The readability of the above code is rather poor. Switch structures should be used for this type of problems, with the syntax of

```
switch switch expression
    case expression 1,statement group 1
```

```
case {expression 2,expression 3,... ,expression m}, statement group 2
    ⋮
otherwise, statement group n
end
```

where the "switch expression" is evaluated first. If its value equals to a condition after a certain case statement, the corresponding statement group should be executed, and then the switch structure is terminated.

When using the switch structures, the following points should be considered:

(1) If the "switch expression" equals "expression 1", the "statement group 1" is executed, and then the switch structure is completed. This structure is different from the syntax in C language, where a break statement should be placed before the next case statement.

(2) If one of the several expressions is met, the expressions must be expressed as a cell array in MATLAB.

(3) If none the expressions enumerated in the case statements are satisfied, the statement group in the otherwise statement are executed. The keyword otherwise here is equivalent to the default keyword in C language.

(4) The result of the program is independent of the orders of the case statements, if there are no repeated expressions.

(5) Repeated expressions should not be used in different case statements, otherwise those listed behind will never be executed.

Switch and conditional structures are essentially conditional structures. If certain conditions are satisfied, the program can be transferred to a corresponding part to continue execution. What are the major differences between the two structures? It can be seen that inequalities can be used in the conditions in the conditional structures, such as if $x>0$, and variable $x$ is regarded as a continuous one. In switch structures, only enumerable samples can be used as the expressions. Therefore, the switch variables are regarded as discrete. It can be seen that the application fields of the two structures are different.

**Example 4.22.** Write a MATLAB program to compute the perimeter and area of a circle, and the volume of a sphere.

**Solutions.** Assumed that the radius is given by $r$. The following appointment can be made for a variable key: if it equals to 1 or 2, the perimeter or area of the circle can be computed, while if it is 3, the volume of the sphere can be computed. Before the execution of the program, the variables $r$ and key should be assigned. Then the program computes, as required, the perimeter, area, and volume, respectively.

```
>> key=menu('Choose a task','perimeter','area','volume');
   r=input('Enter the radius');
```

```
switch key
   case 1, S=2*pi*r;
   case 2, S=pi*r^2;
   case 3, S=4*pi*r^3/3;
end
```

With the above statements, a menu with three options is given, and then a prompt is given to ask for the radius. If "area" is selected from the menu, and the radius is set to 5, the area obtained is $S = 78.5398$.

## 4.4 Trial structure

Trial structure is a special structure provided in MATLAB, with the syntax

`try,` statement group 1, `catch,` statement group 2, `end`

In the trial structure, the statements in "statement group 1" are executed first. If an error occurs in the execution, an error message appears and is sent to the reserved `lasterr`. The execution of the statement group 1 is terminated, and the execution of statements in "statement group 2" is started. If there is no error in the execution in "statement group 1", the whole structure is completed. The statements in "statement group 2" is then not executed.

Trial structure is useful in practical programming. If the solution of a certain problem has two algorithms, one of them is reliable, but is extremely slow, while the other, being not very reliable, is fast, the fast algorithm is embedded in the `try` clause, and the reliable one is placed in the `catch` clause. This programming strategy may make the function reliable and efficient. An alternative application of such a structure is that a certain algorithm may fail due to some reasons, and it can be placed in the `try` clause, the statements in the `catch` clause explain why an error happened. Besides, the trial structure can be used to handle error traps in programming.

**Example 4.23.** There are two ways in MATLAB to represent a number $a$. One method is the double precision scheme, while the other us the symbolic scheme. What we are expecting is that a MATLAB function can be written to test whether a variable $a$ is a number or not. The existing MATLAB function `isnumeric($a$)` can be used to check whether $a$ is a number. However, if $a$ is defined as `sqrt(sym(2))`, then function `isnumeric()` returns 0, which is not what we are expecting. Design a MATLAB function to check if the input variable is a number or not.

**Solutions.** To solve the problem, a MATLAB function should be written. Details about the function will be shown in Chapter 5. If the input argument $a$ is a vector, not a scalar, the function returns 0. If $a$ is a scalar, the data type should be measured. For double precision variables, `key` returns 1. If $a$ is a symbolic variable, we can try to convert it into a double precision variable. If it is successful, *a* is a numeric symbolic variable, double() function call can be placed in the try clause, if there is an error, the code in the catch clause can be executed. Therefore, key can be set to 0, otherwise, key is 1.

With such considerations, the following MATLAB function can be written:

```
function key=isanumber(a)
key=0; if length(a)~=1, return; end
switch class(a)
   case 'double', key=1;
   case 'sym', try, double(a); key=1; catch, end
end
```

The key step in the function is the double() statement for the symbolic variable conversion. If *a* is a symbolic scalar, but not a number, then double() function may lead to errors. The try command is terminated, and the statements in the catch clause are executed. In this clause, the variable key remains unchanged, which is 0, indicating that *a* is not a numeric scalar.

## 4.5 Problems

4.1 Generate a $100 \times 100$ magic matrix. Find all the values larger than 1 000 and set them to zero. Two methods can be used, one involves a loop structure, and the other uses the vectorization method.

4.2 Find all prime numbers smaller than 1 000 with a low-level loop structure.

4.3 Function $A$=rand(3,4,5,6,7,8,9,10,11) can be used to generate a multidimensional array. Find out how many elements are there in $A$, and what is the mean value of all the elements.

4.4 Compute $S = \sum_{i=0}^{63} 2^i = 1 + 2 + 4 + 8 + \cdots + 2^{62} + 2^{63}$ with the numerical method. If a loop structure is not used, compute the sum with the vectorization method. Due to the restrictions in the double format, the result obtained may not be accurate. Compute the accurate result under the symbolic framework.

4.5 The first row and column elements in an *n*th order Pascal matrix are all 1's. The remaining elements of the matrix can be computed from

$$p_{i,j} = p_{i,j-1} + p_{i-1,j}, \quad i = 2, 3, \ldots, n, \ j = 2, 3, \ldots, n.$$

Write a low-level MATLAB command to generate the Pascal matrix. Function pascal() can be used to generate the same matrix, and validate the results and efficiency.

4.6 For a given order $n$, input the following matrix into MATLAB:

$$A = \begin{bmatrix} 1 & -2 & 4 & \cdots & (-2)^{n-1} \\ 0 & 1 & -2 & \cdots & (-2)^{n-2} \\ 0 & 0 & 1 & \cdots & (-2)^{n-3} \\ \vdots & \vdots & \vdots & \ddots & \vdots \\ 0 & 0 & 0 & \cdots & 1 \end{bmatrix}.$$

4.7 An iterative sequence is given by $x_{n+1} = x_n/2 + 3/(2x_n)$, with $x_0 = 1$. If $n$ is large enough, the general term may approach a fixed constant. Select a suitable $n$, compute the steady state value of the terms, with error tolerance of $10^{-14}$. Also find the exact mathematical representation of the result.

4.8 For a given iterative formula $x_{k+1} = (x_k + 2/x_k)/2$, select arbitrarily an initial value of $x_0$, and assign a stop condition to terminate the iterative process. Find out what the limit value of the iterative formula is.

4.9 Compute in loops and vectorized form $S = \prod_{n=1}^{\infty} (1 + 2/n^2)$; the error tolerance can be selected as $\epsilon = 10^{-12}$.

4.10 Expand the polynomial $\prod_{k=1}^{10} (x^k + 2k)$.

4.11 Assume that the general term in the product formula is given by $a_k = (x + k)^{(-1)^k}$, compute the product $a_1 a_2 \cdots a_{40}$.

4.12 Expand the Fibonacci-like sequence, and compute its first 300 terms, where $T(n) = T(n-1) + T(n-2) + T(n-3)$, $n = 4, 5, \ldots$, and the initial values are $T(1) = T(2) = T(3) = 1$.

4.13 Simulate the coin-tossing experiment in Example 4.17, with low-level MATLAB commands.

4.14 Judge whether the problem in Example 4.18 can be computed with vectorized programming method. Why?

4.15 Lagrange interpolation is a commonly used interpolation algorithm[22]. For the given samples $x_i$ and $y_i$, the interpolation of the function at vector $x$ can be computed from

$$\phi(x) = \sum_{i=1}^{m} y_i \prod_{j=1, j \neq i}^{m} \frac{x - x_j}{(x_i - x_j)}.$$

Write a piece of MATLAB code to implement Lagrange interpolation. Hint: the following code is given for reader's own experience.

```
>> ii=1:length(x0); y=zeros(size(x));    % generate initial vectors
   for i=ii, ij=find(ii=i); y1=1;        % remove the current value
       for j=1:length(ij), y1=y1.*(x-x0(ij(j))); end % product
       y=y+y1*y0(i)/prod(x0(i)-x0(ij));   % compute
   end
```

4.16 Monte Carlo method is a commonly used statistical testing algorithm. Consider that a set of $N$ uniformly distributed pseudorandom numbers, inside the square with side length of 1, is given. If $N$ is large enough, the value of $\pi$ can be approximated as $\pi \approx 4N_1/N$, where $N_1$ is the number of random points falling inside the quarter circle. Select a suitable $N$, and observe the approximate value of $\pi$ computed. Hint: the reference MATLAB code for Monte Carlo algorithm is

```
>> N=100000; x=rand(1,N); y=rand(1,N);
   i=(x.^2+y.^2)<=1; N1=nnz(i); p=N1/N*4
```

4.17 We can get a random integer square matrix with $A$=randi([$a_m$,$a_M$], $n$). Generate a $4 \times 4$ square matrix whose elements are in the range of $-8 \sim 8$, and its determinant is 1. Is it possible to construct such a complex matrix?

4.18 For an integer with three digits, if the number equals to the sum of the cubes of its three digits, the number is referred to as a narcissus number. Find all the narcissus numbers. Is it possible to find all the narcissus numbers without using loop structures?

# 5 Function programming and debugging

In MATLAB programming, there are two types of source program supported. MATLAB source code appears in ASCII format. One of the coding formats is the M-script program. The program is composed of a series of MATLAB statements, and they are evaluated in sequence, just as the batch files in DOS. The execution of this type of program is simple, one can simply type in the file name under the >> prompt. The data in MATLAB workspace can be accessed directly by M-scripts. The results are returned back to MATLAB workspace. M-scripts are suitable for dealing with small-scale simple computation problems.

MATLAB programs are ASCII text files. Any text editor can be used to edit MATLAB programs, however, it is recommended that the editor provided in MATLAB be used in MATLAB programming. The MATLAB editor is simple and can be used in editing and debugging MATLAB programs. The MATLAB live editor can also be used in MATLAB programs. Live files are no longer ASCII files, and their suffix is `mlx`.

A MATLAB script is one type of program, and is presented in Section 5.1. The limitations of MATLAB script and the necessity of using MATLAB functions are also addressed. In Section 5.2, the structure of MATLAB functions is proposed, and MATLAB function programming is demonstrated with examples. In Section 5.3, MATLAB programming skills are discussed, including recursive methods, handling of arbitrary number of inputs and returned arguments is presented. Also, global variables and MATLAB workspace assessment are discussed. In Section 5.4, MATLAB function debugging is presented and the debugging facilities are demonstrated with examples. Pseudocode techniques in MATLAB functions are addressed. Live editor and live MATLAB programs are presented in Section 5.5. Demonstrative live MATLAB files are presented through examples.

## 5.1 MATLAB scripts

In this section, MATLAB scripts are discussed. The execution of a MATLAB script is very simple. Just type in the file name at the MATLAB prompt >> and the statements in the MATLAB script are executed automatically. The variables in MATLAB workspace can be processed with MATLAB commands, and the results and intermediate variables are also returned in MATLAB workspace. A MATLAB script is usually suitable for small-scale problems where the answers are immediately expected.

**Example 5.1.** Consider the problem in Example 4.6. Find the solution with MATLAB scripts.

**Solutions.** Typing `edit` at MATLAB command window, or clicking the New Script button in MATLAB toolbar, a MATLAB editor interface can be opened. The code in Example 4.6 can be copied into the editor, as shown in Figure 5.1. This program can be con-

https://doi.org/10.1515/9783110666953-005

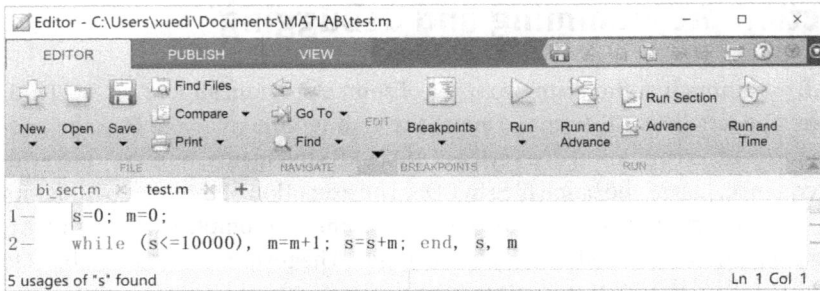

**Figure 5.1:** Program editor interface.

sidered the first MATLAB program in this book. The program can be saved into a file test.m. MATLAB scripts are pure text files.

When the program is established, it can then be executed at any time in MATLAB environment, just by typing in the file name test at the MATLAB prompt. Each time the program is executed, the results are exactly the same.

**Example 5.2.** Consider again the example studied earlier. If the smallest $m$ is required such that the sum is larger than 20 000, or larger than 30 000 in the original Example 4.6, the user needs to modify the source code of the MATLAB script. A mechanism or a program module is expected, such that when 20 000 or another number is fed to the module, the value of $m$ satisfying the conditions is returned. It is no doubt that the request is reasonable.

MATLAB function is the module we are expecting. In the subsequent sections of the chapter, the programming and applications of MATLAB functions are presented.

## 5.2 Fundamental structures of MATLAB functions

A MATLAB function, also known as an M function, is the major structure in MATLAB programming. In practical programming, M-script programming is not recommended. In the subsequent sections, MATLAB functions and some tricks in programming are given. The fundamental structure and naming regulations of MATLAB functions are presented in this section, and examples are given to demonstrate the programming of MATLAB functions.

### 5.2.1 Fundamental function structures

A MATLAB function can be regarded as an information processing unit. It accepts a set of input arguments $in_1$, $in_2$, ..., $in_n$ from the caller. In this book series, the input and output variables of functions are referred to as arguments, in contrast to ordinary variables. The input arguments are transferred into the information processing unit. After manipulation in the unit, the results $out_1$, $out_2$, ..., $out_m$ are transferred back to the caller, as returned arguments. This process is illustrated in Figure 5.2.

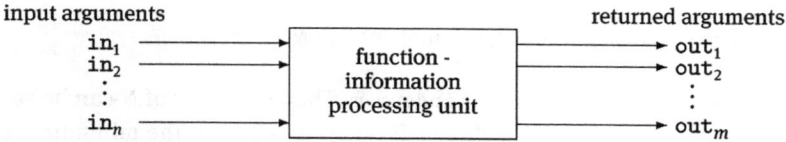

input arguments

$in_1$
$in_2$
$\vdots$
$in_n$

function - information processing unit

returned arguments

$out_1$
$out_2$
$\vdots$
$out_m$

**Figure 5.2:** Illustration of a function.

MATLAB functions are started by the keyword `function`, and the fundamental structure of M function is as follows:

`function` [returned argument list]=function name(input argument list)
comments to the function begin with percentage sign %
checking the input and returned arguments
main body of the function

The actual number of input and returned arguments can be measured with commands `nargin` and `nargout`. Once the function is executed, the two variables can be generated automatically.

If more than one returned argument is used, they should be enclosed with square brackets, otherwise the square brackets can be omitted. The arguments in the lists should be separated with commas. Each line of comments should begin with %, and the characters after the % sign are not executed. The messages in the leading comments can be displayed by the `help` command. Beside, in formal programming, it is necessary to check the number and data types of the input and returned arguments. If the formats of the arguments are incorrect, prompts should be given.

From the system viewpoint, the MATLAB functions acts as an information processing unit, which receives arguments from the caller. Once the variables are processed, the resulted arguments will be sent back to the caller. Apart from the input and returned arguments, the other variables within the function are the local ones, which vanish after function calls. Examples will be given to demonstrate the programming techniques.

**Example 5.3.** An example of a function has been given in Example 4.23, where the input argument is $a$. If $a$ is a scalar and it is numeric, the returned argument key is 1, otherwise, it returns 0.

**Example 5.4.** Consider the problem in Example 5.2. Rewrite it in the form of an M function.

**Solutions.** Consider the requests in Example 5.2. One may choose the input argument as $N$, and returned arguments of $m$ and $s$, where $s$ is the sum of the first $m$ terms. The function can then be written as

```
function [m,s]=findsum(N)   % script can be encapsulated as an M function
s=0; m=0;
while (s<=N), m=m+1; s=s+m; end % original code in M script
```

Such a function can be saved into the file findsum.m. Then the value of $N$ can be selected freely to run the function. For instance, if one wants to find the minimum $m$ such that the sum is larger than 145 323, the following statements can be used, and the results are $m_1 = 539$, $s_1 = 145\,530$.

```
>> [m1,s1]=findsum(145323)  % functions are more flexible
```

It can be seen that the format of a function is more flexible, since there is no need to modify the source code, when the condition is changed. It is recommended to write MATLAB programs in functions.

## 5.2.2 Regulations in function names

The naming regulation of functions is exactly the same as that for variable names, where the first character must be a letter. In real programming, meaningful names should be selected:

(1) Avoid the use of overly simple function names, such as a, otherwise it may conflict with given variables.

(2) Before naming, make sure there are no homonyms under the current MATLAB search path, otherwise, the existing functions with the same name may be shadowed, and unpredictable results may occur.

How can we confirm the nonexistence of homonyms? For instance, if you want to use the name my_fname, you should first execute the command which my_fname, and see whether it returns any information. If no information is displayed, you can safely use the name. Also, the function exist() can be used to find reliable results.

(3) The function names are now case sensitive.

## 5.2.3 Examples of function programming

Some examples are given in this section to show how to write MATLAB functions. The MATLAB scripts can be converted into MATLAB functions, with suggestions on the procedures given here. We also describe how to let MATLAB functions have different syntaxes.

**Example 5.5.** Consider the iterative algorithm for $\pi$, discussed in Example 4.8. Select input and returned arguments and convert the MATLAB script into a MATLAB function.

**Solutions.** Before converting a script into a function, two things must be done first. One is to design the input and returned arguments for the function, and the other is to select a meaningful name. For this particular example, the input argument eps0 can be selected as the error tolerance $\epsilon$, while the returned variable $P$ can be selected as the approximation of $\pi$. Besides, the number of iterations, $k$, can also be selected.

A function name should be selected meaningful, and different from existing ones, such as pi_iter. With such preparations, the original script file can be converted to a function, as shown below.

```
function [P,k]=pi_iter(eps0)
P=1; d=2; d0=sqrt(2); k=1;
while(abs(d-1)>eps0),
    d=d0/2; P=P*d; d0=sqrt(2+d0); k=k+1;
end, P=2/P;
```

The rewritten function can now be called freely in MATLAB environment. For some different error tolerances $\epsilon$, the function can be called directly, and different results can be obtained. If $\epsilon$ is small enough, due to the limitations of the double precision data type, the accuracy of the final result cannot be further increased, even though $\epsilon$ is reduced.

```
>> [p1,k1]=pi_iter(1e-10), [p2,k2]=pi_iter(eps)
   [p3,k3]=pi_iter(1e-20)
```

In actual MATLAB programming, sometimes it is necessary to recognize the number of input and returned arguments. The command nargin can be used to extract the number of input arguments, and nargout command can be used to extract the number of returned arguments. The following example is used to demonstrate the application of these commands.

**Example 5.6.** Suppose one wants to write a MATLAB function to generate an $n \times m$ Hilbert matrix, whose general term is $h_{i,j} = 1/(i+j-1)$. Besides, the following additional considerations are to be kept in mind while implementing the function:

(1) If only one input argument $n$ is given, a square matrix can be generated, i. e., assign automatically $m = n$.

(2) Necessary help information should be included, such as general information, as well as the syntaxes of the function.

(3) Measure the number of the input and returned arguments and see whether there are problems. If there are, error messages should be given.

**Solutions.** In normal cases, the expected numbers of rows and columns, $n$ and $m$, should be given, and double loops should be used to generate the Hilbert matrix from the given general term.

Considering the request in (1), the number of input arguments should be measured with `nargin`, if it reads 1, then let $m = n$, so as to generate a square matrix. With (2), comments should be written, including what the function is, and what the syntaxes of the function are. In actual programming, it is better to write adequate comments, which are beneficial to the programmer as well as to the maintainer of the program. According to (3), the input and output arguments should be checked.

Based on the above considerations, a MATLAB function `myhilb()` can be written, and saved as `myhilb.m` file, in MATLAB search path.

```
function A=myhilb(n,m)
%MYHILB  this function is used to demonstrate MATLAB function programming
%        A=myhilb(n,m) generates an n by m Hilbert matrix A
%        A=myhilb(n) generates an n by n square Hilbert matrix A
%See also: HILB

%Designed by Professor Dingyü Xue, (c) 1995-2017
if nargout>1, error('Too many output arguments.'); end
if nargin==1, m=n; %if only one input argument n is given, let m = n
elseif nargin==0 | nargin>2,
   error('Wrong number of input arguments.'); end
for i=1:n, for j=1:m, A(i,j)=1/(i+j-1); end, end %generate the matrix
```

In the function, the sentences beginning with % are comments. In the first group of the comments, usually the use of the function is described in the first line. This sentence can be searched with `lookfor` function. Then follow the syntaxes of the function, with the descriptions of the input and output arguments. Finally, in the last line, `See also` keywords can be used to list the related MATLAB functions.

The following command can be used to get the help information:

```
>> help myhilb    %   display on-line help information
```

the on-line help information is displayed as follows:

```
MYHILB  this function is used to demonstrate MATLAB function programming
        A=myhilb(n,m) generates an n by m Hilbert matrix A
        A=myhilb(n) generates an n by n square Hilbert matrix A
See also: HILB
```

Note that the information and syntaxes of the function are displayed, while the comment lines with author information are not displayed, since there is a blank line in the source code. It is easily concluded that the `help` command can only display the information in the first few comment lines. If the information in the comments is not to be displayed, there should be a blank line in front of it.

With the function established, the following different syntaxes can be used and the expected results can be obtained:

```
>> A1=myhilb(4,3), A2=myhilb(sym(4)) % yield different matrices
```

The resulted matrices are

$$
A_1 = \begin{bmatrix} 1 & 0.5 & 0.33333 \\ 0.5 & 0.33333 & 0.25 \\ 0.33333 & 0.25 & 0.2 \\ 0.25 & 0.2 & 0.16667 \end{bmatrix}, \quad A_2 = \begin{bmatrix} 1 & 1/2 & 1/3 & 1/4 \\ 1/2 & 1/3 & 1/4 & 1/5 \\ 1/3 & 1/4 & 1/5 & 1/6 \\ 1/4 & 1/5 & 1/6 & 1/7 \end{bmatrix}.
$$

**Example 5.7.** Let us consider again Example 5.5. If there is no input argument given in the function, a default error tolerance eps can be assumed. Modify the original MATLAB function to implement the assignment of the default input arguments.

**Solutions.** If the number of input arguments is zero, which means that the value of `nargin` is 0, the following new function can be written:

```
function [P,k]=pi_iter1(eps0)
if nargin==0, eps0=eps; end
P=1; d=2; d0=sqrt(2); k=1;
while(abs(d-1)>eps0),
    d=d0/2; P=P*d; d0=sqrt(2+d0); k=k+1;
end, P=2/P;
```

**Example 5.8.** Consider the bisection algorithm implementation in Example 4.16. Convert the MATLAB scripts into an M function.

**Solutions.** It is seen from the original problem that the input arguments can be selected as $f$, where $f$ can be an anonymous function or other function handles. Apart from $f$, the input arguments should also include $a$, $b$, and the error tolerance `err`. The returned arguments are $x$, the solution, $k$, the number of iterations, and `f1`, the value of the function when the solution is substituted. In this way, the M script file can easily be converted into an M function.

```
function [x,k,f1]=bi_sect(f,a,b,err)
k=0;
while abs(b-a)>err, x=(a+b)/2; k=k+1;
```

```
    if f(x)*f(a)<0, b=x; else, a=x; end
end
```

With such an M function, if one wants to solve the algebraic equation $f(x) =$ $x^2 \sin(0.1x + 2) - 3 = 0$, an anonymous function should be created. Also, the user has to select $a$ and $b$, such that $f(a)$ and $f(b)$ are of different signs. Otherwise, the bisection algorithm cannot be used. The following statements can be used to solve the above equation. It can be seen that the function format is much more convenient and flexible than the use of M scripts.

```
>> f=@(x)x.^2.*sin(0.1*x)-3; [x,k]=bi_sect(f,-4,5,1e-12)
```

## 5.3 Skills of MATLAB function programming

Apart from the regular programming essentials, in order to improve the efficiency of the functions, some skills in MATLAB programming are also important. In this section, recursive structure of functions is demonstrated, and the limitations of the structure are all pointed out. Besides, discussions are made on how to accept arbitrarily many input arguments in functions. The topics of global and private variables are also presented. Also, the workspace assessment to and from functions is discussed, and special functions such as anonymous functions, subfunctions, private functions are addressed.

### 5.3.1 Recursive structures

**Definition 5.1.** A function in MATLAB may call itself, and this type of function is referred to as a recursive function.

**Example 5.9.** Write a factorial $n!$ computation function with a recursive structure.

**Solutions.** Considering the example of the factorial $n!$, a very useful recursive formula $n! = n(n - 1)!$ can be used. Therefore, a MATLAB function my_fact() can be written, with the kernel statement

$k=n*$my_fact$(n - 1)$

where $k$ is the expected factorial $n!$. This relationship is referred to as a recursive relationship.

It is obvious that the recursive statement is not adequate for implementing the function, since the relationship will be infinitely executed. In order to correctly compute the factorial, an exit should be written, to let the recursive process terminate. It is known that the factorial of $n$ can be evaluated from the factorial of $n - 1$, which can be further evaluated from the factorial $n - 2$, and so on, until $1! = 0! = 1$. The conditions $1! = 0! = 1$ can be used as the exit of the recursive process. The comments in the function are omitted.

```
function k=my_fact(n)
if nargin~=1, error('Error: Only one input variable accepted'); end
if abs(n-floor(n))>eps | n<0     % check whether n is a nonnegative integer
    error('n should be a non-negative integer'); % display error message
end
if n>1, k=n*my_fact(n-1);         % if n > 1, use recursive formula
elseif any([0 1]==n), k=1; end % 0! = 1! = 1, the recursive process exits
```

It can be seen that the function tests whether $n$ is a nonnegative integer first. If it is not, there is no factorial, so an error message is given to terminate the function. If it is, then the recursive formula is implemented when $n > 1$, to call itself. If $n = 1$ or $0$, the function returns 1 directly. With function my_fact(11), the results of the function is $11! = 39\,916\,800$.

In fact, a factorial function, factorial() is provided in MATLAB, with the kernel algorithm of prod(1:n), and the structure is simpler, more straightforward, and faster.

**Example 5.10.** Try to compute the Fibonacci sequence with a recursive structure, and compare it with loop structures.

**Solutions.** Recursive structure is an effective algorithm for a class of problems, however, it should not be misused in actual programming. A counterexample is considered here. Consider the recursive formula of the Fibonacci sequence, with the first two terms $a_1 = a_2 = 1$. From the $k$th term on $(k = 3, 4, \dots)$ the term can be computed recursively from $a_k = a_{k-1} + a_{k-2}$. It is quite natural to use a recursive algorithm to implement the algorithm. The exits of the function are at $k = 1, 2$, with an output of 1. The listing of the function is given by

```
function a=my_fibo(k)   % function with recursive structure
if k==1 | k==2, a=1;
else, a=my_fibo(k-1)+my_fibo(k-2); end
```

In the function, the statement to test whether the input argument $k$ is a positive integer is omitted. If the 40th term is expected, the following statements can be written, and the time needed is 13.2 seconds. With earlier versions, it took much longer.

```
>> tic, my_fibo(40), toc % the 40th term, only one term is returned
```

When using the function with the recursive algorithm, if the 42nd term is expected, the time needed is 41.02 seconds, while when $k = 50$, it may take several days to find the results. What is the reason for such a large computational burden? Assume the term required is $k = 8$, then the terms when $k = 7$ and $k = 6$ should be computed again from the very beginning, while the computation of $k = 7$ term also needs to compute terms with $k = 6$ and $k = 5$ from the very beginning. A tree structure is used to compute the terms. Therefore, the computational load is extremely heavy. If the information of terms of $k = 7$ and $k = 6$ can be reused for $k = 8$, the computational load will be reduced significantly.

Loop structure can be used to compute all the terms when $k \leqslant 100$, and the total elapsed time is only 0.0002 seconds.

```
>> tic, a=[1,1];
   for k=3:100, a(k)=a(k-1)+a(k-2); end, toc % first 100 terms
```

It can be seen that loop structure is suitable for this kind of problems. Therefore, recursive structures should not be misused. Further observing the results, it can be seen that the above double precision scheme is not adequate to find the accurate results of the whole sequence. If symbolic data type is used, the first statement should be changed from `a=[1,1]` into `a=sym([1,1])`. It can be seen that the 100th term is $a_{100} = 354\,224\,848\,179\,261\,915\,075$, with total elapsed time of 0.8 seconds.

```
>> tic, a=sym([1,1]); for k=3:100, a(k)=a(k-1)+a(k-2); end, toc
```

### 5.3.2 Functions with variable numbers of inputs and outputs

One of the most important applications of cell arrays is presented below. We are interested in how to establish a syntax of a function which allows arbitrary numbers of input and returned arguments. It should be pointed out that many MATLAB functions are using this method in dealing with input and returned arguments.

Before introducing the method for manipulating arbitrary numbers of input and returned arguments, the method of passing the input arguments from the caller is presented first. Based on the argument transferring mechanism in MATLAB, the input arguments are passed to the function in a cell array, named `verargin`. The structure is illustrated in Figure 5.3, where $n$ can be measured with `nargin`. Similarly, the returned arguments are passed to the caller with a cell array named `varargout`.

**Example 5.11.** There are two ways of representing a polynomial in MATLAB. One is with a symbolic expression, and the other is with the coefficient vector, in descending

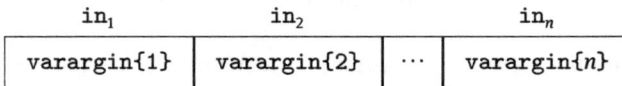

| $\text{in}_1$ | $\text{in}_2$ | | $\text{in}_n$ |
|---|---|---|---|
| varargin{1} | varargin{2} | ⋯ | varargin{$n$} |

**Figure 5.3:** The passing pattern of the input arguments.

order of the independent variable $s$. The latter case is considered here in this example. Function conv() in MATLAB can be used to multiply two polynomials together. When one tries to multiply more polynomials together, the nested function could be used, and it is rather complicated. Write a MATLAB function, so that an arbitrary number of polynomials can be multiplied together with a single function call.

**Solutions.** The target name of the function can be selected as convs(). The input arguments can be represented in the cell array varargin, while the returned argument is the vector $a$, containing the coefficients in the resulted polynomial. The initial value of $a$ can be set to 1, then a loop structure can be used to extract each input argument and multiply it to $a$. The input arguments can be extracted as shown in Figure 5.3, and conv() can be used to multiply the input argument with $a$. The final result can be returned in vector $a$.

```
function a=convs(varargin)
a=1; % set the initial value
for i=1:nargin,
   a=conv(a,varargin{i}); % step extracts an input argument and multiplies
end
```

All the input arguments are represented in the cell array varargin, with the $i$th input argument stored in varargin{$i$}. Similarly, the returned arguments are stored in the cell array varargout. Theoretically speaking, any number of input and returned arguments can be processed with the structures. The function can be called with the following statements:

```
>> P=[1 2 4 0 5]; Q=[1 2]; F=[1 2 3];
   D=convs(P,Q,F)        % the product of 3 polynomials
   E=conv(conv(P,Q),F) % nested calls of the conv() function
   G=convs(P,Q,F,[1,1],[1,3],[1,1]) % any number of input arguments
```

The identical $D$ and $E$ vectors can be found, also the vector $G$ is

$$D = [1, 6, 19, 36, 45, 44, 35, 30]^T,$$

$$G = [1, 11, 56, 176, 376, 578, 678, 648, 527, 315, 90]^T.$$

**Example 5.12.** An interface to MuPAD was studied in Example 2.20. Now, a MATLAB function can be written from which we can learn the programming of the interfaces.

**Solutions.** The listing given in Example 2.20 is written here again

```
function p=padefrac(f,varargin)
[x,n,m]=default_vals({symvar(f),2,2},varargin{:});
```

```
orders=['[' int2str(n) ',' int2str(m) ']'];
p=feval(symengine,'pade',f,x,orders);
```

In the first statement, a fixed input argument $f$ is used. Other input arguments are passed in the `varargin`. A low-level function `default_vars()` can be written to extract the default values

```
function varargout=default_vals(vals,varargin)
if nargout~=length(vals), error('number of arguments mismatch');
else, nn=length(varargin)+1;
    varargout=varargin; for i=nn:nargout, varargout{i}=vals{i};
end, end, end
```

where `vals` is the list of default values, provided as a cell array. With the low-level function, if function `padefrac()` is called with `padefrac(f)`, `varargin` in function `default_vals()` is empty, and then the three defaults values are used. If the function call is `padefrac(f,x)`, the latter two default values are assigned.

### 5.3.3 Fault tolerance manipulation

Although different syntaxes are allowed in MATLAB functions, the support of the syntaxes should be programmed by the user. In practical applications, function `nargchk()` can also be used to check the numbers of input and returned arguments. If the numbers are not correct, the function can be terminated automatically, and error messages are given:

```
msg=nargchk(nm,nM,'nargin' or 'nargout')
```

where $n_m$ is the minimum allowed number of arguments, and $n_M$ is the maximum. The last input argument given here is to indicate which arguments are checked. The returned argument `msg` is a string, which stores the error messages. If the arguments are checked, `msg` returns an empty string.

**Example 5.13.** Consider again the polynomial product function considered in Example 5.11. In normal cases, each polynomial is given by a row vector. If one accidentally specifies one or several column vectors, direct use of the function may give wrong results. Fault tolerance programming in the source function can be introduced.

**Solutions.** The simplest way for fault tolerant programming is that, when the $i$th vector is extracted, no matter whether it is a row or column vector, we can use ( : ) to uniformly convert it into a column vector, and then with . ', all the vectors are converted into the expected row vectors. The modified MATLAB function can then be obtained.

```
function a=convs1(varargin)
a=1; % set the initial value
for i=1:nargin,
    a=conv(a,varargin{i}(:).'); % convert the argument to a row vector
end
```

### 5.3.4 Global variables

All the variables shared by the MATLAB function and its caller are passed by input and returned arguments. The other variables inside a MATLAB function are all local variables. They will vanish automatically when the function call is terminated. Although the same variable names are used inside the function and in MATLAB workspace, they are mutually independent. The local variables are not affecting any variable in MATLAB workspace.

**Example 5.14.** Consider the MATLAB function in Example 5.5. The variable eps0 is an input argument, $P$ and $k$ are the returned arguments. The variables $d_0$ and $d$ inside the function are local variables. When the function call is completed, they will automatically vanish.

In practical MATLAB programming, if one wants to have two or more functions share the same variables, it is necessary to introduce the concept of global variables. Global variables can be declared with the command global. They should be declared at least in two places. One is in the program the value of the variable is assigned, while the other is where the variable is used or modified, inside a function.

**Example 5.15.** Consider the function designed in Example 5.8. If the variables $a$ and $b$ are represented by global variables, modify the function.

**Solutions.** If the parameters $a$ and $b$ are assigned as global variables, it is not necessary to have them in input arguments. Internally, a command global can be used to directly declare global variables. The original function can then be modified as

```
function [x,k,f1]=bi_sect1(f,err)
global a b, k=0;
while abs(b-a)>err, x=(a+b)/2; k=k+1;
    if f(x)*f(a)<0, b=x; else, a=x; end
end
```

To solve the equation, the variables $a$ and $b$ should also be declared as global variables, and their values should be assigned. Then the following MATLAB statements can be used to solve the equation, and the same results can be obtained:

```
>> global a b; a=-4; b=5;
   f=@(x)x.^2.*sin(0.1*x)-3; [x,k]=bi_sect1(f,1e-12)
```

It can be seen that the global variable data transfer is more complicated than that with input and returned arguments. If it is not absolutely necessary, global variables are not recommended.

### 5.3.5 Reading and writing of MATLAB workspace

In certain applications, the results internally obtained inside a MATLAB function are required to be written directly to MATLAB workspace. While some other times, the MATLAB workspace variables need to be loaded into the function. In this section, the data exchange between MATLAB function and MATLAB workspace are presented. Global variables can be used to solve this kind of problems. Better methods are introduced here.

The pair of MATLAB functions, `assignin()` and `evalin()` can be used to transfer variables between a function and MATLAB workspace. With function `assignin('base',varname,var)`, the variable var inside a function can be written to MATLAB workspace, with a variable name of varname. Also, the function $a$=`evalin('base',var)` can be used to load MATLAB workspace variable var into a function, with the name $a$.

The option 'base' can also be substituted into 'caller', such that the data transfer is not between the function and MATLAB workspace, it is between the function and its caller.

**Example 5.16.** Consider the function designed in Example 5.15, where the global variables $a$ and $b$ were used. Use `evalin()` instead of global variables, and read the variables $a$ and $b$ from MATLAB workspace directly. Rewrite the MATLAB function.

**Solutions.** If `global` command is not used, function `evalin()` can be used to load variables $a$ and $b$ from MATLAB workspace. The original MATLAB function can then be rewritten as

```
function [x,k,f1]=bi_sect2(f,err)
k=0; a=evalin('base','a'); b=evalin('base','b');
while abs(b-a)>err, x=(a+b)/2; k=k+1;
    if f(x)*f(a)<0, b=x; else, a=x; end
end
```

If one wants to solve the equation, the variables $a$ and $b$ must be assigned in MATLAB workspace, then the function can be used to solve the original equation. The result obtained is exactly the same as that in the previous examples.

```
>> a=-4; b=5; % assign variables in MATLAB workspace
   f=@(x)x.^2.*sin(0.1*x)-3; [x,k]=bi_sect2(f,1e-12)
```

**Example 5.17.** An objective function was designed in [21]. Inside the function, the decision variable vector $x$ is written based on variables $K_p$, $K_i$, and $K_d$ in MATLAB workspace. The subsequent Simulink model can be used to utilize the parameters. The data transfer like this can be implemented with function `assignin()`. In the function, the following code can be written:

```
assignin('base','Kp',x(1)); assignin('base','Ki',x(2));
assignin('base','Kd',x(3));
```

Therefore, in each run of the function, the decision variable vector $x$ can be written to MATLAB workspace, to the expected variables.

It should be noted that the variable transfer methods are irregular, formal methods with input and returned arguments should be used instead.

### 5.3.6 Anonymous and inline functions

So far, the MATLAB functions we discussed were those from a *.m file. Sometimes, for the convenience of simple function descriptions, anonymous functions can be used. The syntax of an anonymous function is similar to that of ordinary MATLAB functions, and only one returned argument is allowed, which is referred to as the function handle. There is no need to write a genuine *.m file. With such a MATLAB function, the mathematical function can be dynamically defined. The syntax of anonymous function is

$f$=@(input arguments) function expression,
e. g., $f$=@$(x,y)$sin$(x.$^2+$y.$^2$)$

The existing variables in MATLAB workspace can be used directly in anonymous functions. For instance, if there are two variables $a$ and $b$ in MATLAB workspace, the following anonymous function can be defined:

$f$=@$(x,y)a*x.$^2+$b*y.$^2$,
whose mathematical form is $f(x,y) = ax^2 + by^2$

There is no need to arrange $a$ and $b$ in the input argument list, therefore, the description of a mathematical function is made more convenient. Note that, when the anonymous function is defined, the current values of $a$ and $b$ are used. If the variables $a$ and $b$ are later altered, their values used in the anonymous functions are not changed. If one wants to use the new values of $a$ and $b$ in the anonymous function, the anonymous function should be defined again. One must be very careful when the variables in MATLAB workspace are used.

Anonymous functions have been used many times earlier. They are very useful in numerical computation. A simple mathematical function can be expressed with anonymous functions easily.

**Example 5.18.** When $a = 1$ and $b = 2$, define an anonymous function for $f(x, y) = ax^2 + by^2$, and compute $f(2, 3)$. If $a = 2$ and $b = 1$, what is $f(2, 3)$?

**Solutions.** The values of $a$ and $b$ are assigned first, then an anonymous function can be defined. The value of the anonymous function can be obtained as $f(2, 3) = 22$. The anonymous function is essentially defined as $f=@(x,y)x^2+2*y^2$. If the variables $a$ and $b$ are changed, the anonymous function is not changed at all. Therefore we still have $f(2, 3) = 22$, not the value of 17 in the redefined $f_1$.

```
>> a=1; b=2; f=@(x,y)a*x^2+b*y^2; f(2,3)
   a=2; b=1; f(2,3), f1=a*2^2+b*3^2
```

Why does such a situation happen? In defining the anonymous function, the variables $a$ and $b$ are not actually used. What are used are the current values of the variables. After the anonymous functions are defined, even though the values of the variables $a$ and $b$ change, the anonymous functions are not changed.

Is one wants the anonymous function to change with the values of $a$ and $b$, the variables should be used as the input arguments of the anonymous function. Or, when $a$ and $b$ are changed, the anonymous function should be defined again to force the change.

```
>> f=@(x,y)a*x^2+b*y^2; f(2,3)
```

In the earlier versions of MATLAB, `inline()` functions are also used. The facilities of `inline()` functions are very similar to the anonymous functions, however, the variables in MATLAB workspace cannot be used. Besides, it is less effective. The syntax of the function is `fun=inline(function expression,input arguments)`, where "function expression" is given in a string, and "input arguments" are a series of independent variable names given as strings. For instance, function $f(x, y) = \sin(x^2 + y^2)$ can be described with

$f$=inline('sin($x$.^2+$y$.^2)','$x$','$y$')

It can be seen that anonymous function descriptions are much simpler and more flexible. The use of `inline()` function is not recommended. It is advised that anonymous functions are used for describing simple mathematical functions.

Anonymous and `inline()` functions may only return one variable $f$, known as a function handle. Compared with regular MATLAB functions, the limitations of these functions are that intermediate statements cannot be used in them. The advantages of these functions are that the descriptions are simpler, also, no *.m files are required.

### 5.3.7 Subfunctions and private functions

If in a function A a low-level function B is called, and function B is only called by function A, with no other function calling it, function B can be appended to function A, in the same file. There is no need to make a separate *.m file for function B. In this case, function B is referred to as a subfunction of function A. Subfunctions are only visible to their main functions. Other functions cannot call subfunction B.

The name of a subfunction can be assigned arbitrarily, as long as it starts with a letter. Even though the name of the subfunction coincides with other function names, its priority is higher than that of other functions. The other functions with the same name are not affected.

**Example 5.19.** Consider the function in Example 5.8. Rewrite the function such that the kernel part is expressed as a subfunction.

**Solutions.** A subfunction abc() is written, which is the kernel part of the original function. It can be appended in the same file. The new MATLAB function can be obtained from

```
function [x,k,f1]=bi_sect3(f,a,b,err)
k=0; while abs(b-a)>err, [a,b,x]=abc(f,a,b); k=k+1; end
function [a,b,x]=abc(f,a,b)
x=(a+b)/2; if f(x)*f(a)<0, b=x; else, a=x; end
```

Since the subfunction abc() is only called once in the main function, there is actually no need to use the subfunction format. The function in Example 5.8 is sufficient for this problem. This example is only used for demonstrating the use of subfunctions.

Another category of special functions in MATLAB are private functions. In folder A, a low-level private folder can be created. Some functions can be placed in this folder as private functions, such that MATLAB in other folders cannot see the private functions. The private functions can only be called by the functions in folder A. The information of the private functions cannot be displayed with command which and lookfor.

Apart from the special format of functions, overload functions can also be written in MATLAB. Overload functions should be placed in the folders starting with @ in their names. They belong to the corresponding classes. Overload functions will be presented further in Chapter 9, where object-oriented programming is presented.

## 5.4 MATLAB function debugging

Comprehensive debugging facilities are provided in MATLAB. Breakpoints can be set in functions such that the internal local variables can be assessed. Also, the internal statements of the function can be executed in the single-step mode. The debugging facilities can be used in an interface, as well as by debugging commands. In this section, the debugging facilities are demonstrated through examples. Pseudocode techniques and applications in MATLAB are also presented.

### 5.4.1 Debugging of MATLAB functions

As mentioned earlier, the information transfer between the caller and the function is accomplished through input and returned arguments. Apart from the arguments, other variables are local variables, which cannot be monitored outside the function. Debugging facilities can be introduced to monitor the changes of local variables, when the function is executed.

Similar to other computer languages, breakpoints can be set in a MATLAB function. Here, an example is presented to demonstrate the debugging facilities in MATLAB editor interface.

**Example 5.20.** Consider the bisection solver in Example 5.8. The debugging facilities can be demonstrated here.

**Solutions.** The MATLAB code in Example 5.8 is listed here again.

```
function [x,k,f1]=bi_sect(f,a,b,err)
k=0;
while abs(b-a)>err, x=(a+b)/2; k=k+1;
    if f(x)*f(a)<0, b=x; else, a=x; end
end
```

In the loop, the value of $x$ is computed in each step, and the values of $a$ or $b$ are updated. However, since this information is stored in the intermediate local variables, which are invisible to the caller, debugging facilities can be used to monitor the changes in the variables. For instance, a breakpoint, indicated by a red filled circle, can be made at the statement if, as shown in Figure 5.4. To set a breakpoint, one should click in the editor the "−" sign in front of the corresponding line. If one clicks this circle again, the breakpoint can be removed. If necessary, several breakpoints can be set by the user.

Having defined the breakpoint, the following statement can be executed:

```
>> f=@(x)x.^2.*sin(0.1*x)-3; [x,k]=bi_sect(f,-4,5,1e-12)
```

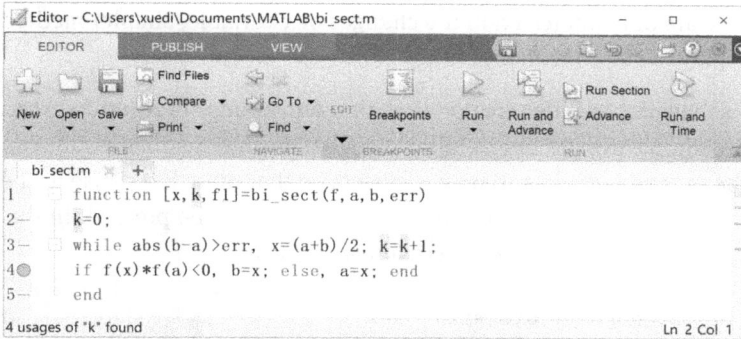

**Figure 5.4:** Setting of a breakpoint.

The program will stop at the breakpoint, and in MATLAB command window, the prompt is changed to K>>, indicating that the internal local variables can be assessed at this point. For instance, if one wants to display the local variable $x$, the value of it is 0.5. There are two ways to monitor a local variable. One way is to type x at the K>> prompt, while the other is to move (not click) the cursor over the x variable in the editor, the value of the variable can be displayed automatically, as shown in Figure 5.5. The other local variables can be traced in a similar way.

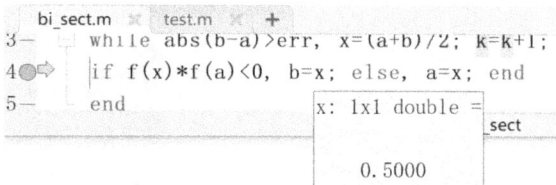

**Figure 5.5:** Display of an internal local variable.

If one clicks the **Step** button in the toolbar of the editor, the program will execute one statement and stop again. One may click **Continue** to execute the program to the next breakpoint. In this case, if $x$ is monitored again, the value of $x$ is displayed as $x = 2.75$. Further **Step** execution may further display $x = 3.875$, $x = 3.3125$, ... In this way, the internal local variable can be monitored.

If the current function calls a lower-level function, the button **Step In** can be clicked such that lower-level functions can be monitored in a similar manner. To return to the upper-level function, **Step Out** button can be clicked.

If one wants to execute the program to the end, the breakpoint can be cleared first, and then one should click **Continue** button, or press **Quit Debugging** button. In this case, the MATLAB prompt is changed back to >>.

**Example 5.21.** If we are only interested in the changes of *x* variable, simpler method can be used to display the intermediate results.

**Solutions.** If the debugging facilities are not used, the source code of the function can be used directly, by adding an extra line as follows, so that the intermediate results can be obtained. In the function, command pause is used, so that there is a pause in the execution. To resume the execution of the function, any key can be pressed. In this manner, the intermediate result of the *k*th iteration is displayed.

```
function [x,k,f1]=bi_sect3(f,a,b,err), k=0;
while abs(b-a)>err, x=(a+b)/2; k=k+1;
    disp(['when k=' int2str(k),', x=' num2str(x,16)]), pause
    if f(x)*f(a)<0, b=x; else, a=x; end
end
```

The following commands can be used to execute the program:

```
>> f=@(x)x.^2.*sin(0.1*x)-3; [x,k]=bi_sect3(f,-4,5,1e-15);
```

and the intermediate results can be displayed as follows:

```
    when k=1, x=0.5
    when k=2, x=2.75
    when k=3, x=3.875
    . . .
    when k=51, x=3.12418273039707
    when k=52, x=3.124182730397072
    when k=53, x=3.124182730397073
```

**Example 5.22.** Considering the function bi_sect3() and other similar ones, if the error tolerance selected is too small, the code will be trapped in an infinite loop. How can you modify the code such that the function can be terminated normally?

**Solutions.** If one selects err as $10^{-19}$, the conditions in the while loop will never be satisfied, therefore, a lower bound should be set to err. An appropriate one can be 2×eps. The original function should be modified as

```
function [x,k,f1]=bi_sect4(f,a,b,err), k=0;
while abs(b-a)>max(err,2*eps), x=(a+b)/2; k=k+1;
    disp(['when k=' int2str(k),', x=' num2str(x,16)]),
    if f(x)*f(a)<0, b=x; else, a=x; end
end
```

### 5.4.2 Pseudocode and code protection

The objectives of using MATLAB pseudocode are as follows:

(1) The speed of the function can be increased. By using the pseudocode technique, MATLAB code can be converted to a somewhat executable code, and the conversion is no longer needed, such that the total execution time is reduced. Since the conversion process in MATLAB is relatively fast, normally the increase is not significant. When complicated graphical user interfaces are used, the speed of a pseudocode may be significantly increased.

(2) Pseudocode can be used to convert ASCII MATLAB source code into binary code, so that other users can execute the code normally, but are unable to read the source code, so as to protect the source code.

The `pcode` command provided in MATLAB can be used to do the conversion, and the suffix of the converted files is .p. If one wants to convert a MATLAB function mytest.m to P code, the command to use is `pcode mytest`; if one wants the generated P code located in the same folder of the M code, the command

```
pcode mytest -inplace
```

can be used; if one wants to convert all the .m files in a folder into P code, he/she can enter the folder with cd command, then type the command `pcode *.m`. If there is no error in the source code, all the files can be converted, otherwise, the conversion will be aborted, and error messages will be displayed. The programmer can also use this method to check whether his/her source code has syntax errors or not. If both *.m and *.p exist in the same folder, the *.p file has higher priority in execution.

Please note that the user must save the original *.m files in safe places. Make sure that the source files are not deleted, otherwise, they cannot be recovered from the *.p code.

## 5.5 MATLAB live editor

Live editor is recently introduced in MATLAB. The facilities of live editor are quite similar to those of a Mathematica notebook, or to a notebook (M-book) in the earlier versions of MATLAB. M-book was a Microsoft Word based editor, which was not quite convenient to use.

Live editor can be used to edit enhanced text file with graphics and executable MATLAB commands. Executable MATLAB statements and programs can be embedded in the file. While browsing the text, the embedded programs can also be executed. The suffix of a live document is *.mlx, rather than *.m in the standard MATLAB editors. The live document is no longer ASCII.

### 5.5.1 Live editor interface

An ordinary MATLAB editor can be used to generate a blank live document. Clicking the File → New → Live File menu item, a blank live document can be opened, and the editor itself is switched into Live mode, as shown in Figure 5.6. The manipulation of live documents is allowed in the new editor.

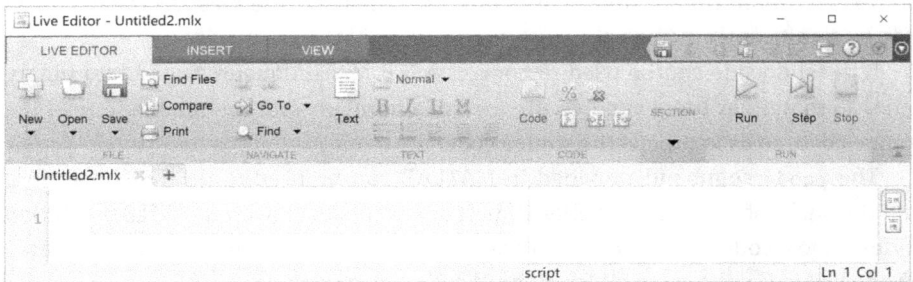

**Figure 5.6:** Live editor interface.

Live files are similar to an electronic document, where the user can write an article with it. The title and author information of the article can be provided, and the article has a regular structure. For instance, **Sections** structure can be used. Besides, formulas, figures, and links are all supported. The most significant feature is that executable MATLAB code can be embedded, such that it can be executed live to compute online the results or new figures. The parameters or statements can be modified to get different results.

Each line of the text in live editor has two modes, text and code, which can be tangled with Text or Code buttons in the toolbar. It can be seen from the interface that the code can be filled in the shaded box, while text is not. If the text mode is selected, the buttons in the toolbar are provided, which allows the user to select the modes of the text, such as bold, italic, the icons are given just as in Microsoft Word editor. The user may feel it easy to build text and other objects in the live editor.

### 5.5.2 Creating a live document

Clicking the black triangle by the Normal icon, a listbox is opened with several options: Normal, Heading 1, Heading 2, and Title. A text format can be selected by the user. An example is given below to show the facilities.

**Example 5.23.** Use live editor to create a live document as shown in Figure 5.7.

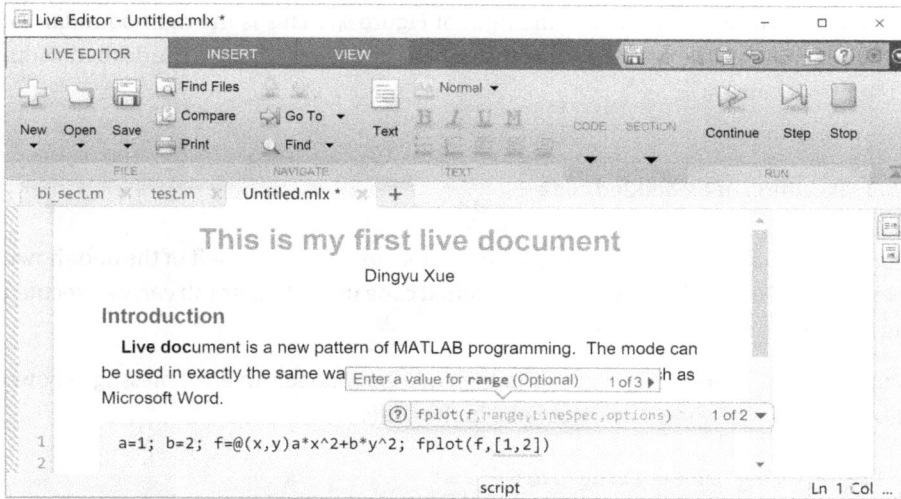

**Figure 5.7:** Live editor interface.

**Solutions.** A blank live document can be opened first with live editor. Then the following guidelines are used to create the expected live document:

(1) Select Text mode. Select from Normal listbox the Title format, and type in This is my first live document. Clicking the center icon ⬛, the text is centered. This completes the display of the first line, with font and color automatically assigned.

(2) In the second line, "Normal" mode is automatically switched to. The author information can be entered and centered.

(3) Changing the Normal mode to Heading 1, then entering "Introduction", the default left aligned mode can be assigned to start a new section. The font and size are all assigned automatically.

(4) The text mode is automatically changed back to Normal. The text in the demonstrative document can be entered directly. Selecting the text Live doc, then clicking B button, the selected text appears in bold. The remaining text is not affected by this action.

(5) Clicking Code icon, a shaded box is given, and MATLAB commands can be entered into the box. Prompts may be displayed on the syntaxes of the functions, and this is rather helpful in writing a MATLAB code. When fplot() is called, the prompts on the syntax is displayed, as shown in Figure 5.7. If the brackets [] are typed, prompt is given to ask you to input the ranges.

With the standard file saving facilities, a live document can be stored as my_live_ed.mlx file.

Note the vertical scroll bar to the right of Figure 5.7. This is the right boundary of the live document. The user may drag the right boundary to adjust to the desired width of the live document.

### 5.5.3 Execution of embedded code

In fact, when a live file is generated, a blue region appears on the left of the one shown in Figure 5.7. Double-clicking it, the embedded code in the paragraph can be executed and the results can be obtained.

**Example 5.24.** There exists an error in the code embedded. An error message shown below indicates that there are problems in the code.

```
Error using fplot (line 112)
Input must be a function or functions of a single variable.
```

Later it will shown that function `fplot()` can only be used to draw plots for explicit functions. Since an implicit function $f(x, y)$ is defined, function `fplot()` cannot be used, `fimplicit()` function should be used instead.

Besides, various other errors are contained in the original function call. For instance, the dot operation should be used, and also $ax^2 + by^2 = 0$ has no curve. It can be modified to $ax^2 + by^2 - 2 = 0$ or other functions. The correct drawing command should be issued as

```
>> a=1; b=2; f=@(x,y)a*x.^2+b*y.^2-2; fimplicit(f,[-2,2])
```

When the newly modified embedded code is executed, an ellipse will be drawn in a new graphics window.

### 5.5.4 Embed other objects in live editor

Various objects can be inserted into a live document. Clicking INSERT tag on the top of the live editor, the options are displayed in Figure 5.8. It can be seen that, apart from Text and Code, Section Breaks, images, hyperlinks, equations, and many others can all be inserted into a live document. The object insertion methods are demonstrated as follows:

(1) Images. Clicking the Image button in Figure 5.8, a standard file name dialog box can be opened, from which an image file can be selected, and inserted into the live document. When an image file is inserted in the document, even if the original file is deleted or modified, the image in the live document is unchanged.

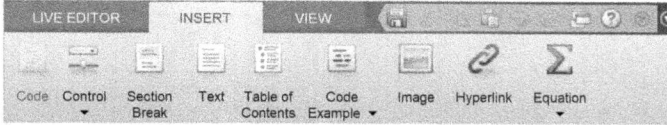

**Figure 5.8:** Objects can be inserted in a live document.

(2) **Equations.** Two ways are supported to insert an equation into a live document. One is to use an equation editor to edit the equation, the other is to use a LaTeX commands to describe the equation. Clicking the Equation button in Figure 5.8, an equation editor similar to that in Microsoft Word is opened, as shown in Figure 5.9. With the equation editor interface, mathematical formulas can be created easily.

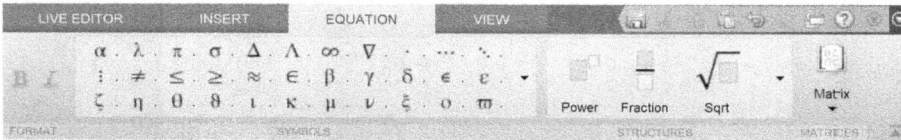

**Figure 5.9:** The toolbar of the equation editor.

Apart from this typical equation editor, clicking the black triangle under the Equation button, LaTeX Equation can be selected from it, and the editing interface in Figure 5.10 is displayed. Standard LaTeX code can be embedded in the live editor. Note that LaTeX commands are not fully compatible in this editor. Only simple LaTeX commands are supported. Complicated ones, as well as those defined by the user, are not supported at all.

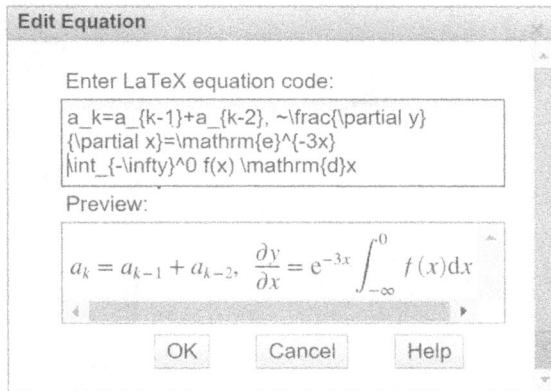

**Figure 5.10:** LaTeX input interface in the equation editor.

**Example 5.25.** Type the following mathematical formula into the live document:

$$a_k = a_{k-1} + a_{k-2}, \quad \frac{\partial y}{\partial x} = e^{-3x} \int\limits_{-\infty}^{0} f(x)\,dx.$$

**Solutions.** The following low-level LaTeX command can be embedded into a live document. The expected preview results are displayed in Figure 5.10.

```
a_k=a_{k-1}+a_{k-2}, ~\frac{\partial y}
{\partial x}=\mathrm{e}^{-3x}
\int_{-\infty}^0 f(x) \mathrm{d}x
```

The LaTeX support in the live editor is not ideal. Although the following simple LaTeX command is used

```
\begin{array}{ccc} 1 & 2 & 3\\ 4 & 5 & 6\\ 7 & 8 & 0\end{array}
```

its formula display is only a mark `\beginarrayccc1`. Double clicking the mark, an equation editor interface can be opened, with the source code. One must be very careful in using the live editor LaTeX equation. For instance, the **Matrix** input format in Figure 5.9 should be considered.

(3) Hyperlinks. Clicking the **Hyperlink** button, the dialog box shown in Figure 5.11 is displayed. One can click the **Display Text** column to all the hyperlink commands. In the **Target URL** box, the commands such as `http://...`, or `mailto:` can be specified.

(4) **Section Break.** Separators can be used to divide a live document into two or more independent sections. The users can then process individual sections of the live document, without affecting the others, by introducing a section break.

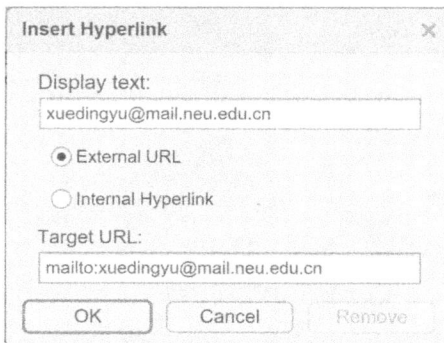

**Figure 5.11:** Hyperlink input interface.

**Example 5.26.** The complete screenshot of a full live document is shown in Figure 5.12. The file is stored in `my_live_ed.mlx`. The readers are advised to open the file, and compare with the screenshot to understand details in live document processing.

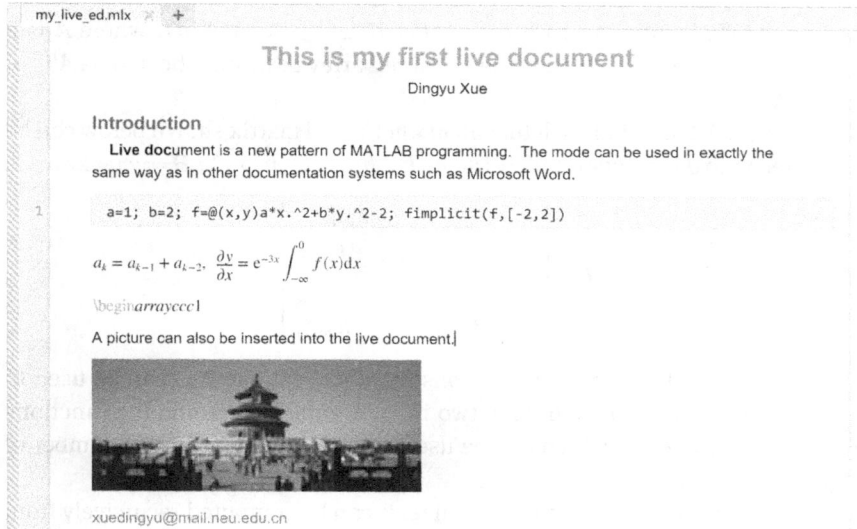

**Figure 5.12:** Screenshot for the generated live document.

### 5.5.5 Output of live files

The default suffix for a live document is `mlx`. Apart from the typical document styles, the documents can also be outputted as PDF, HTML, or LaTeX files. This can be done by selecting the black triangle under the **Save** button. Although there is undisplayed information, such as the unconverted `\beginarrayccc1` in Example 5.25, the converted LaTeX source code or PDF files are correct.

Note that, since the output file has been converted to static text, a live document cannot be restored from the output files.

## 5.6 Problems

5.1   Write a MATLAB function `mat_add()`, which can be used in adding arbitrary numbers of matrices with the format
$$A=\mathtt{mat\_add}(A_1,A_2,A_3,\dots)$$

5.2 Express the following piecewise function in MATLAB:

$$y = f(x) = \begin{cases} h, & x > D, \\ h/Dx, & |x| \leqslant D, \\ -h, & x < -D. \end{cases}$$

5.3 Write a MATLAB function with the syntax $H$=mat_roots$(A,n)$, where $A$ is a square matrix, $n$ is an integer, while $H$ is a cell array returning the $n$ roots $A^{1/n}$ of matrix $A$.

5.4 Write a MATLAB function such that an $m \times m$ Hankel matrix shown below can be generated with the function call $v$= $[h_1, h_2, h_m, h_{m+1}, \ldots, h_{2m-1}]$ ; $H$=myhankel$(v)$.

$$H = \begin{bmatrix} h_1 & h_2 & \cdots & h_m \\ h_2 & h_3 & \cdots & h_{m+1} \\ \vdots & \vdots & \ddots & \vdots \\ h_m & h_{m+1} & \cdots & h_{2m-1} \end{bmatrix}.$$

5.5 The functions gcd() and lcm() demonstrated in Example 3.33 can be used to compute the GCD and LCM of only two input arguments. Extend the functions to gcds() and lcms(), which can be used to deal with an arbitrary number of input arguments.

5.6 For the Fibonacci sequence, the general term can be computed recursively from $a_k = a_{k-1} + a_{k-2}$, $k = 3, 4, \ldots$, with initial terms $a_1 = a_2 = 1$. Write a MATLAB function to compute the Fibonacci sequence, and
(1) with $y$=fib$(k)$, assign the $k$th term to variable $y$;
(2) check the numbers of input and returned arguments;
(3) use recursive structure for the function.

5.7 The following algorithm can be used to approximately compute the value of $\pi$, and it can be implemented with a loop structure:

$$\frac{2}{\pi} \approx \frac{\sqrt{2}}{2} \cdot \frac{\sqrt{2+\sqrt{2}}}{2} \cdot \frac{\sqrt{2+\sqrt{2+\sqrt{2}}}}{2} \cdots$$

When the multiplier $\delta$ satisfies $|\delta - 1| < 10^{-6}$, the loop can be terminated. Is it possible to have a better approximation by reducing the error tolerance of $\pi$? What is the best error tolerance for computing $\pi$ in the double precision framework?

5.8 Newton–Raphson iterative algorithm can be used for solving an equation $f(x) = 0$. Selecting an initial guess of $x_n$, the next approximation can be obtained with $x_{n+1} = x_n - f(x_n)/f'(x_n)$. If the two points are close enough, i. e., $|x_{n+1} - x_n| < \epsilon$, where $\epsilon$ is the error tolerance, find the solution with $x_0 = -4$, and $\epsilon = 10^{-12}$.

5.9 The well-known Mittag-Leffler function is defined as

$$E_\alpha(x) = \sum_{k=0}^{\infty} \frac{x^k}{\Gamma(\alpha k + 1)}$$

where $\Gamma(x)$ is the Gamma function which can be evaluated by gamma(x). Write a MATLAB function with syntax $f$=mymittag($\alpha,z,\epsilon$), where $\epsilon$ is the error tolerance, with default value of $\epsilon = 10^{-6}$. Argument $z$ is a numeric vector. Draw the curves for Mittag-Leffler functions with $\alpha = 1$ and $\alpha = 0.5$.

5.10 Chebyshev polynomials are mathematically defined as

$$T_1(x) = 1, \quad T_2(x) = x, \quad T_n(x) = 2xT_{n-1}(x) - T_{n-2}(x), \quad n = 3,4,5,\dots$$

Write a recursive function to generate a Chebyshev polynomial, and compute $T_{10}(x)$. Write a more efficient function as well to generate Chebyshev polynomials, and find $T_{30}(x)$.

5.11 From matrix theory it is known that if a matrix $M$ is expressed as $M = A + BCB^{\mathrm{T}}$, where $A$, $B$ and $C$ are the matrices of relevant sizes, and the inverse of $M$ can be calculated by the following algorithm:

$$M^{-1} = (A + BCB^{\mathrm{T}})^{-1} = A^{-1} - A^{-1}B(C^{-1} + B^{\mathrm{T}}A^{-1}B)^{-1}B^{\mathrm{T}}A^{-1}.$$

Write the statement to evaluate the inverse matrix. Check the accuracy of the inversion. Compare the accuracy of the inversion method and the direct inversion method with inv() function when

$$M = \begin{bmatrix} -1 & -1 & -1 & 1 & 0 \\ -2 & 0 & 0 & -1 & 0 \\ -6 & -4 & -1 & -1 & -2 \\ -1 & -1 & 0 & 2 & 0 \\ -4 & -3 & -3 & -1 & 3 \end{bmatrix}, \quad A = \begin{bmatrix} 1 & 0 & 0 & 0 & 0 \\ 0 & 3 & 0 & 0 & 0 \\ 0 & 0 & 4 & 0 & 0 \\ 0 & 0 & 0 & 2 & 0 \\ 0 & 0 & 0 & 0 & 4 \end{bmatrix},$$

$$B = \begin{bmatrix} 0 & 1 & 1 & 1 & 1 \\ 0 & 2 & 1 & 0 & 1 \\ 1 & 1 & 1 & 2 & 1 \\ 0 & 1 & 0 & 0 & 1 \\ 1 & 1 & 1 & 1 & 1 \end{bmatrix}, \quad C = \begin{bmatrix} 1 & -1 & 1 & -1 & -1 \\ 1 & -1 & 0 & 0 & -1 \\ 0 & 0 & 0 & 0 & 1 \\ 1 & 0 & -1 & -1 & 0 \\ 0 & 1 & -1 & 0 & 1 \end{bmatrix}.$$

5.12 If the interfaces are designed, many MATLAB scripts in the problems in Chapter 4 can be rewritten as MATLAB functions. For instance, if the order $n$ is given, a MATLAB function can be written to input the following matrix:

$$A = \begin{bmatrix} 1 & -2 & 4 & \cdots & (-2)^{n-1} \\ 0 & 1 & -2 & \cdots & (-2)^{n-2} \\ 0 & 0 & 1 & \cdots & (-2)^{n-3} \\ \vdots & \vdots & \vdots & \ddots & \vdots \\ 0 & 0 & 0 & \cdots & 1 \end{bmatrix}.$$

Write a MATLAB function with the syntax $A$=mymatx($n$). If the input argument $n$ is not given, a $6 \times 6$ matrix can be generated.

5.13 Write a MATLAB function to generate the first $m$ terms of the extended Fibonacci sequence $T(n) = T(n-1) + T(n-2) + T(n-3)$, $n = 4, 5, \ldots$, with $T(1) = T(2) = T(3) = 1$.

5.14 Write a MATLAB function to generate a random integer matrix, with syntax $A$=unirandi([$a_m$,$a_M$],$n$). It is required that the determinant of the matrix is 1. Generate with this function a $13 \times 13$ integer matrix, whose elements can only be selected as 0, 1, and $-1$, and its determinant is 1.

5.15 In an $n$th order Pascal matrix, all the elements in the first row and first column are 1, and the rest of the elements can be computed from

$$p_{i,j} = p_{i,j-1} + p_{i-1,j}, \quad i = 2, 3, \ldots, n, \ j = 2, 3, \ldots, n.$$

Write a MATLAB function to generate the $n$th order Pascal matrix.

5.16 Comparing Problems 4.7 and 4.8 in Chapter 4, it can be seen that one of them is used to compute $\sqrt{2}$, while the other is for $\sqrt{3}$. It can be conceived that if the iterative formula is changed to $x_{k+1} = (x_k + a/x_k)/2$, the formula can be used to compute $\sqrt{a}$. Write a MATLAB function to validate the conjecture.

5.17 Write a short passage to describe the bisection algorithm in solving equations, with the live editor.

# 6 Two-dimensional graphics

Graphical display and visualization is one of the most important features in MATLAB. A series of straightforward two- and three-dimensional graphical facilities are provided in MATLAB. They can be used to display experimental and simulation results in a visible way.

Brand new plotting facilities are provided in MATLAB R2014b, although some of the graphical functions in older versions can still be used, some of them may vanish gradually. In this book series, we are trying to introduce new facilities in MATLAB visualization.

We will concentrate on various two-dimensional graphical facilities. In Section 6.1, fundamental two-dimensional plots for experimental samples and mathematical functions are drawn. Decoration methods for two-dimensional plots are presented in Section 6.2, such that objects like LaTeX formulas and arrows can be added to the plots. Different two-dimensional graphics are presented in Section 6.3, including polar, stem, and stairs plots, histograms, filled and logarithmic plots. Dynamic trajectories and animations of two-dimensional plots are also presented. In Section 6.4, a graphics window partition technique is presented such that the users may draw plots in any portion of the window. Implicit functions can be drawn with the method studied in Section 6.5. In Section 6.6, an introduction to digital image processing is presented. The topics included in the section are: image input, edge direction, and histogram equalization. In Section 6.7, the output of MATLAB graphics is presented.

## 6.1 Drawing two-dimensional plots

Two-dimensional graphs are commonly used in scientific research. They are also the most practical ones. In this section, two-dimensional plots are presented for the given experimental data, then, for given mathematical functions, graphical displays are also presented. Besides, the topics of graph decoration and multiple-axes graphics are presented.

### 6.1.1 Plotting data

Assume that a set of experimental data are acquired. For instance, at time instances $t = t_1, t_2, \ldots, t_n$, the measured function values are $y = y_1, y_2, \ldots, y_n$, then, the data can be input into MATLAB workspace, and they can be stored as vectors $\boldsymbol{t} = [t_1, t_2, \ldots, t_n]$ and $\boldsymbol{y} = [y_1, y_2, \ldots, y_n]$. If the relationship of the two vectors is expected, the command plot$(\boldsymbol{t}, \boldsymbol{y})$ can be applied, to draw a two-dimensional curve. It can be seen that the

https://doi.org/10.1515/9783110666953-006

syntax of the function is straightforward. For practical applications, the extended versions of the syntaxes of plot() are summarized below:

(1) Suppose $t$ is a vector, and $y$ is a matrix given below. The relationship between each row of $y$ and vector $t$ can be drawn, therefore, altogether $m$ plots can be drawn in the same coordinates. Note that the number of columns in $y$ should be the same as the length of $t$.

$$y = \begin{bmatrix} y_{11} & y_{12} & \cdots & y_{1n} \\ y_{21} & y_{22} & \cdots & y_{2n} \\ \vdots & \vdots & \ddots & \vdots \\ y_{m1} & y_{m2} & \cdots & y_{mn} \end{bmatrix}$$

(2) Assuming $t$ and $y$ are both matrices of the same size, the plots between corresponding rows of $t$ and $y$ can be drawn.

(3) Assuming there are many pairs of such vectors or matrices, $(t_1, y_1)$, $(t_2, y_2)$, ..., $(t_m, y_m)$, the following statement can be used directly to draw the corresponding curves:

plot$(t_1, y_1, t_2, y_2, \ldots, t_m, y_m)$

(4) The line type, width and color information of the curves can separately be specified with the command

plot$(t_1, y_1, \text{option } 1, t_2, y_2, \text{option } 2, \ldots, t_m, y_m, \text{option } m)$

where the "options" can be represented as combinations of the strings in Table 6.1.

For instance, if a red dashed–dotted plot is expected, and at each data point, a pentagram is used, the combination of the options can be expressed by the string 'r-.pentagram'.

**Table 6.1:** Different options in MATLAB plots.

| line types | | line colors | | mark symbols | |
|---|---|---|---|---|---|
| option | meaning | option | meaning | option | meaning |
| '-' | solid | 'b' | blue | '*' | star |
| '--' | dashed | 'g' | green | '.' | dots |
| ':' | dotted | 'r' | red | 'x' | cross |
| '-.' | dashed–dotted | 'w' | white | 'v' | ▽ |
| 'none' | none | 'c' | cyan | '^' | △ |
| | | 'k' | black | '>' | ▷ |
| | | 'm' | magenta | 'pentagram' | ☆ |
| | | 'y' | yellow | 'o' | circles |
| | | | | 'square' | □ |
| | | | | 'diamond' | ◇ |
| | | | | 'hexagram' | ✩ |
| | | | | '<' | ◁ |

(5) Command $h$=plot$(\cdots)$ can be used such that the handle of the plots is returned in $h$. Later, the handles can be used to point to the plots, such that the properties of the plots can be extracted or modified.

After the plots are drawn, the command grid on can be used to add mesh grids to the axis, or grid off can be applied to remove the mesh grids. Besides, command hold on can be used to preserve the axes, such that the subsequent plot() function call may superimpose curves on top of the original ones. The command hold off can be used to cancel the hold mode.

Compared with the powerful plot() function, a low-level function line() can be used to superimpose the curves on the existing ones, where hold on is not necessary. No handle is allowed to return with such a function.

**Example 6.1.** Draw the curve of the explicit function $y = \sin(\tan x) - \tan(\sin x)$ in the interval $x \in [-\pi, \pi]$.

**Solutions.** The following direct commands can be tried to draw the plot:

```
>> x=[-pi : 0.05: pi];           % create a vector with step-size of 0.05
   y=sin(tan(x))-tan(sin(x)); plot(x,y) % compute and draw the plot
```

With the above commands, the graphics window is opened automatically as shown in Figure 6.1. It seems that the curve obtained is suspicious. In the graphics window, if one moves the mouse mark over the axis, a toolbar above the axis is displayed automatically.

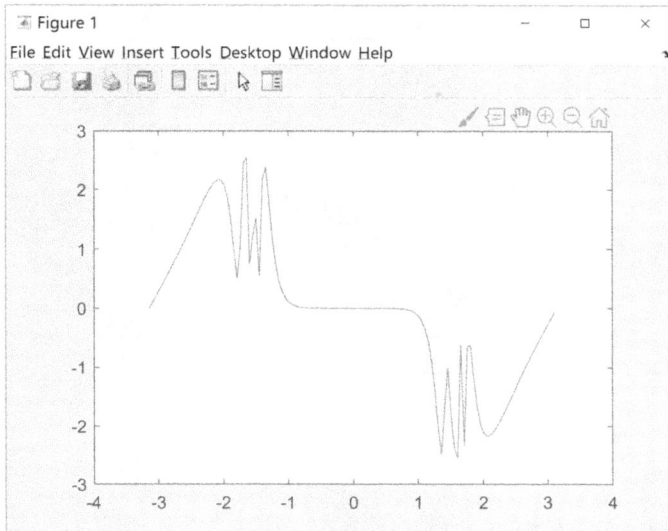

**Figure 6.1:** Graphical representation of the given function.

It should be noted that the curves drawn with MATLAB function `plot()` are not really curves. They are just segments connecting the samples in the data. If the samples are densely distributed, the segments look like curves. Later in this series, the plots are still referred to as "curves".

**Example 6.2.** Generate densely distributed samples, draw the correct curve for the function in Example 6.1.

**Solutions.** Observing the curves shown in Figure 6.1, it can be seen that the plot around the points $\pm\pi/2$ is suspicious, while elsewhere the curve is smooth. Why does this happen? Observing the $\sin(\tan x)$ term, since at the points around $\pm\pi/2$, $\tan x$ function tends to infinity, the value of $\sin(\tan x)$ is irregular and unpredictable, so it is oscillatory at these points.

A small step size over the whole range of $x$ can be used. Alternatively, a small step size can be used in the two subintervals $x \in (-1.8, -1.2)$ and $x \in (1.2, 1.8)$, and in the remaining parts, the original step size can be used. Therefore, the following statements can be used instead:

```
>> x=[-pi:0.05:-1.8,-1.799:0.001:-1.2, -1.2:0.05:1.2,...
        1.201:0.001:1.8, 1.81:0.05:pi];  % variable step-size selected
    y=sin(tan(x))-tan(sin(x)); plot(x,y) % compute and draw plot
```

The new plot is obtained as shown in Figure 6.2. It can be seen that the new plot looks reliable, since it has been explained earlier that there exist strong oscillations in neighborhoods of points $\pm\pi/2$.

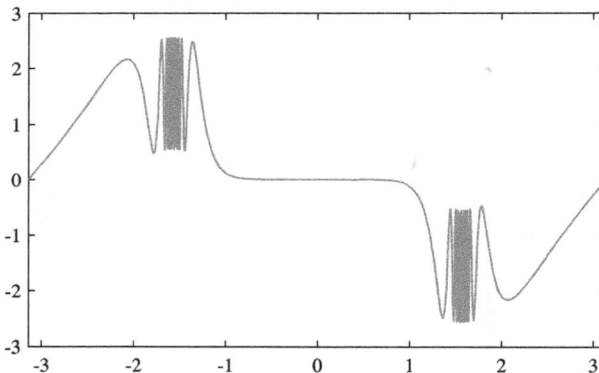

**Figure 6.2:** Plot with a smaller step size.

Note that, in this and subsequent plots, the graphics windows are no longer provided. Only the plots in the window are shown in the series.

As suggested by this example, we should not trust too much the plots generated automatically in MATLAB. Normally, the correctness of the curves obtained should be validated. A feasible solution is that a smaller step size can be used and a new plot can be obtained. The two curves should be compared to see whether they look the same. If the two curves match well, the curves obtained are regarded as correct, otherwise, even smaller step sizes should be tried, until validated results can be obtained.

### 6.1.2 Plots of mathematical functions

Apart from the data-based `plot()` function, if the mathematical function is known, function `fplot()` can be used to draw the function directly, with `fplot(f)`, where $f$ can be an anonymous function, a symbolic expression, or a symbolic function. The default range for the plot is $[-5, 5]$. If one wants to specify the range, the command `fplot(f, [x_m, x_M])` can be used. In earlier versions of MATLAB, function `ezplot()` can be used instead, with the default range of $[-2\pi, 2\pi]$.

**Example 6.3.** Draw the plot of the function in Example 6.1 again with `fplot()` function.

**Solutions.** A symbolic function can be used to describe the original mathematical function. The plot can be obtained with the following MATLAB statements. It can be seen that the result is not as good as that where samples are selected by the user.

```
>> syms x; f(x)=sin(tan(x))-tan(sin(x)); fplot(f,[-pi,pi])
```

The original function can also be described by an anonymous function, and identical results can be obtained.

```
>> f=@(x)sin(tan(x))-tan(sin(x)); fplot(f,[-pi,pi])
```

### 6.1.3 Plots of piecewise functions

A vectorized programming technique was discussed in Section 4.2.3. Now it can be used to compute piecewise functions. Also the anonymous functions presented in Chapter 5 can be used. With these methods, the functions `plot()` or `fplot()` can be used to draw directly the piecewise functions. In this section, an example is used to demonstrate the plotting facilities for piecewise functions.

**Example 6.4.** Draw the following saturation nonlinearity:

$$y = \begin{cases} 1.1\,\text{sign}(x), & |x| > 1.1, \\ x, & |x| \leqslant 1.1. \end{cases}$$

**Solutions.** Three methods can be used in presenting piecewise functions: (1) vectorized computing, (2) anonymous function expression, and (3) symbolic expression. The plots of these three methods are demonstrated in this example.

Let us see method (1). If vectorized method is used, the following statements can be used to evaluate the piecewise function, and finally, the plot can be obtained, as shown in Figure 6.3.

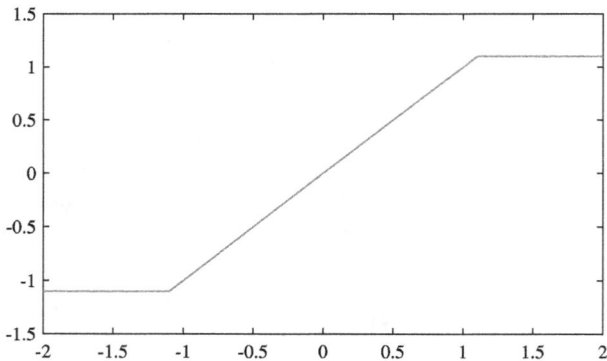

**Figure 6.3:** Plot of the piecewise function.

```
>> x=[-2:0.02:2]; y=1.1*sign(x).*(abs(x)>1.1)+x.*(abs(x)<=1.1);
   plot(x,y)
```

It should be noted that, in describing the piecewise function, the conditions should be mutually exclusive. For instance, the first condition should not be expressed as $x>=1.1$, otherwise there exists an overlap between the two conditions at $x_i = 1.1$. If $x_i = 1.1$ is a repeated point, then erroneous result can be obtained.

Besides, since plot() function can only be used to join the samples together with segments, the finite turning points for this function can be extracted, and the simple command given below can be used to draw the plot. The result is exactly the same as that in Figure 6.3.

```
>> plot([-2,-1.1,1.1,2],[-1.1,-1.1,1.1,1.1]) % join the turning points
```

Now let us see method (2). An anonymous function can be used to represent the original piecewise function, then fplot() function can be used to draw the plot, which is exactly the same as that obtained earlier.

```
>> f=@(x)1.1*sign(x).*(abs(x)>1.1)+x.*(abs(x)<=1.1);
   fplot(f,[-2,2]), ylim([-1.5 1.5]) % setting manually the vertical axis
```

Method (3) can also be used, and the result is identical.

```
>> syms x; clear f % clear is necessary here to remove the anonymous function
   f(x)=piecewise(abs(x)>1.1,sign(x),abs(x)<=1.1,x);
   fplot(f,[-2,2]), ylim([-1.5 1.5]) % setting manually the vertical axis
```

### 6.1.4 Titles in plots

In the plots generated in MATLAB, each curve is an object. The axis and graphics window are both objects, with different properties. The properties can be modified with function set(), also they can be extracted with the function get(). The syntaxes of the functions are

set(handle,'prop name 1',prop value 1,'prop name 2',prop value 2,...),
$v$=get(handle,'property name')

Besides, function title() can be used to add a title to a plot. Functions xlabel() and ylabel() can be used to add titles to $x$ and $y$ axes, respectively. Strings can be provided to these functions as input argument. The font, its size, and text rotations can be made automatically by the functions.

**Example 6.5.** Add a title and axes titles to the plot obtained in Example 6.4.

**Solutions.** The plot can be redrawn, and then functions like title() can be called to decorate the plots, as shown in Figure 6.4. The fonts and their sizes can be assigned to these titles automatically to the plot. Also, the $y$ axis label can be rotated by 90°. Characters in other languages can also be added, however, when *.eps files are generated, these characters cannot be processed correctly in the graphics files. Better solutions to these problems should be introduced.

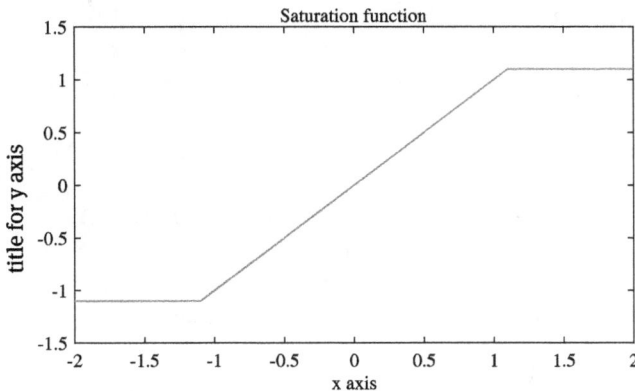

**Figure 6.4:** The saturation plot with titles added.

```
>> plot([-2,-1.1,1.1,2],[-1.1,-1.1,1.1,1.1]) % saturation function
   title('Saturation function'),  % add the title
   xlabel('x axis'), ylabel('title for y axis')
```

Functions such as xlim() and ylim() can be used to assign the ranges for the axes. The syntaxes of these functions are very simple, for instance, xlim([$x_m$,$x_M$]) command can be used. No further example is given for that.

**Example 6.6.** Consider the fractional tree model in Example 4.18. Draw the fractional tree for the points denoted by dots.

**Solutions.** The data for the fractional tree can be generated directly with the commands in Example 4.18. With the data generated, two vectors **x** and **y** can be constructed. With the data, the fractional tree shown in Figure 6.5 can be obtained.

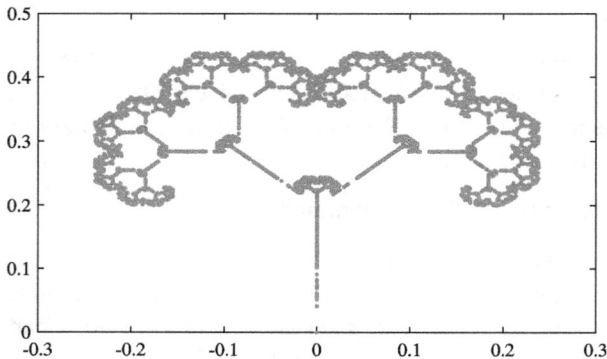

**Figure 6.5:** Fractional tree display.

```
>> v=rand(10000,1); N=length(v); x=0; y=0;
   for k=2:N, gam=v(k);
       if gam<0.05, x(k)=0; y(k)=0.5*y(k-1);
       elseif gam<0.45,
           x(k)=0.42*(x(k-1)-y(k-1)); y(k)=0.2+0.42*(x(k-1)+y(k-1));
       elseif gam<0.85,
           x(k)=0.42*(x(k-1)+y(k-1)); y(k)=0.2-0.42*(x(k-1)-y(k-1));
       else, x(k)=0.1*x(k-1); y(k)=0.1*y(k-1)+0.2;
   end, end
   plot(x,y,'.')   % note that, the dot option should not be omitted
```

### 6.1.5 Plots with multiple vertical axes

For two sets of data, if their values differ a lot, although they can be displayed under the same coordinates, the readability of the plot with smaller magnitudes may not be good. A new function `plotyy()` can be used for better display, with the syntax `plotyy(`$x_1$`,`$y_1$`,`$x_2$`,`$y_2$`)`. An example given below can be used to demonstrate the plotting facilities.

**Example 6.7.** Consider two functions, $y_1 = \sin x$ and $y_2 = 0.01 \cos x$. Show them in the same coordinates.

**Solutions.** The following statements can be used directly to show the two curves, as shown in Figure 6.6. Since the two curves differ greatly in magnitude, the readability of $y_2$ is very poor, and its change can hardly be noticed from the curves. Therefore, this method is not suitable for representing such curves.

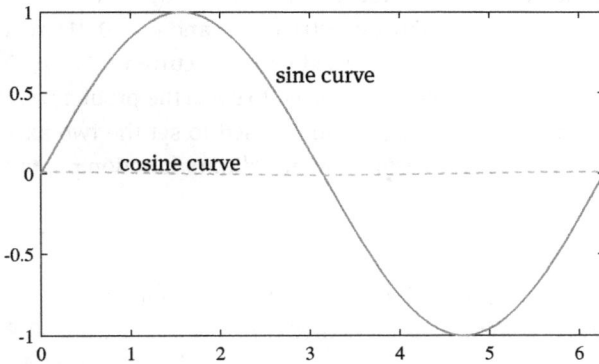

**Figure 6.6:** Plots of two functions with different magnitudes.

```
>> x=0:0.01:2*pi; y1=sin(x); y2=0.01*cos(x);
   plot(x,y1,x,y2,'--') % the plots in the same coordinates
```

For this kind of problems, a suitable function is `plotyy()`, where plots with two $y$ axes can be drawn as shown in Figure 6.7.

```
>> plotyy(x,y1,x,y2)   % double y-axis plot can be obtained
```

**Example 6.8.** The plots with two vertical axes can be handled individually. For instance, command `yyaxis` can be used to assign a certain axis as the current one, then individual processing of the axes is made possible. Manipulate similar problems with this method.

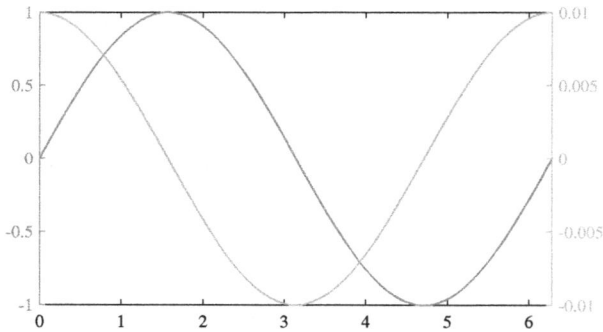

**Figure 6.7:** Plots with two vertical axes.

**Solutions.** Only two plots can be drawn with the current function `plotyy()`. If more than two curves are expected, they can be assigned to different axes. Now consider an application as follows. In the first axis, the function $y_1 = \sin x$ can be drawn, while in the other axis (the axis on the right), two functions $y_2 = 0.01\cos x$ and $y_3 = 0.01\sin 2x$ are expected. This kind of problem cannot be processed with the current `plotyy()` function. Individual axis manipulation method can be used to solve the problem.

The commands `yyaxis left` or `yyaxis right` can be used to set the two axes individually to bring the left or right axis to the front. Then, ordinary function `plot()` can be used to draw directly the plots.

```
>> t=linspace(0,2*pi,500);
   yyaxis left; plot(t,sin(t)) % assign the left axis as the current one
   yyaxis right, plot(t,0.01*cos(t),t,0.01*sin(2*t),'--')
```

In certain applications, plots with three or even four vertical axes can be drawn by special functions (see Figure 6.8) `plotyyy()`[7] and `plot4y()`[3], downloadable for

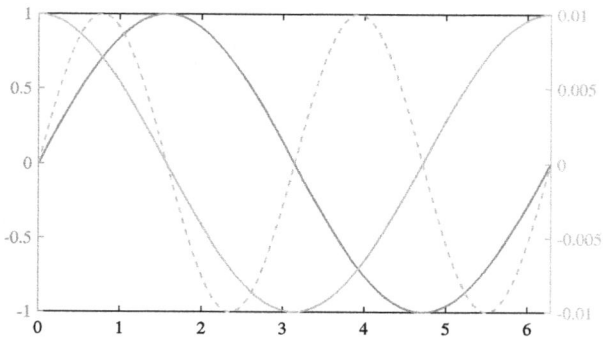

**Figure 6.8:** The plots of three functions.

free from the MathWorks File Exchange website. With function `plotxx()`, the plots with double *x* axes can be obtained[6].

## 6.2 Decoration of plots

Plot decoration facilities are provided in MATLAB. If the menu items View → Figure Palette, View → Plot Browser, View → Property Edit Toolbar and View → Property Editor are all selected, the edit window shown in Figure 6.9 can be obtained, while in its toolbar, the users are allowed to add arrows, text, double arrows, ovals, and boxes. These objects enhance the capabilities of MATLAB graphical decorations. Besides, zooming and three-dimensional rotation of the plots are also allowed.

### 6.2.1 Plot decoration with interface tools

Three major portions of the graphics window are usually involved in graphics editing. The left portion corresponds to the View → Property Edit Toolbar menu allowing the

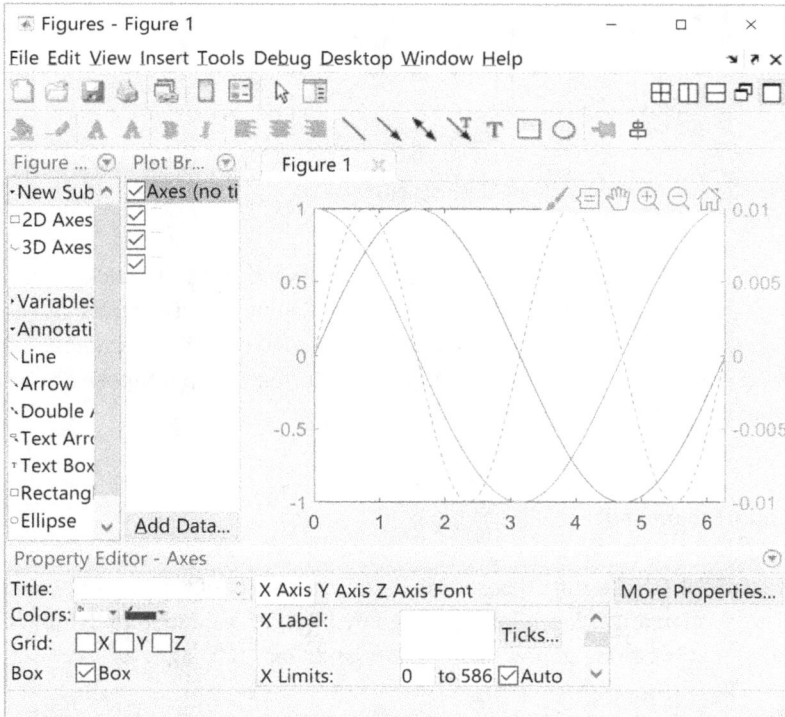

**Figure 6.9:** MATLAB graphical editing interface.

user to add the decorations such as arrows, text or ovals to the graphics window. Two-an three-dimensional coordinates can be added. The bottom portion corresponds to the **Property Editor**, which allows the user to select the properties in colors, fonts, and line types of the objects. The portion on the window corresponds to View → Plot Browser, allowing the user to select elements in the window to edit them. If one adds new data, the new plots are superimposed on the existing ones.

In the toolbar above the axis, a button ▤ can be used to read the information on the curves with mouse. It is suitable for the implementations of graphical methods in scientific computing. Even though three-dimensional plots are shown, the facilities can still be used to read coordinate information of the selected points. If one clicks the button ◉, the two-dimensional plots can be displayed in three-dimensional plots, as shown in Figure 6.10.

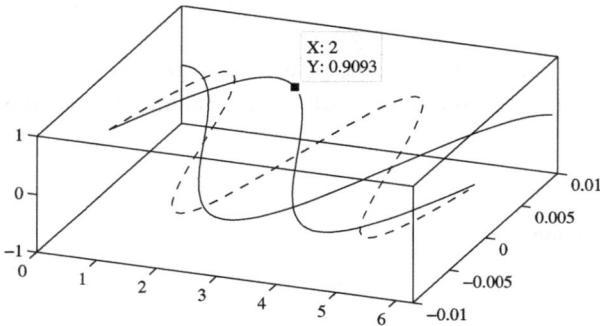

**Figure 6.10:** Three-dimensional display of two-dimensional plots.

If the text box labeled T is clicked from the toolbar, the position can be picked with the mouse, such that strings can be specified directly to display text. The string can be composed of letters and text, and LaTeX format can be used to describe mathematical formulas. Also the tools like arrows and lines can be selected to superimpose arrows and lines on the plots.

### 6.2.2 LaTeX support commands

LaTeX is a well-established scientific typesetting system. A subset of LaTeX is supported in MATLAB. The mathematical formulas described by LaTeX can be added to the plots.

(1) Special symbols can be generated by commands led by \ sign, as shown in Table 6.2.

(2) Superscript and subscript signs can be represented with ^ and _. For instance, a_2^2+b_2^2=c_2^2 represents $a_2^2 + b_2^2 = c_2^2$. If many symbols are used as a superscript,

**Table 6.2:** Supported LaTeX commands in graphics windows.

| type | sign | LaTeX | sign | LaTeX | sign | LaTeX | sign | LaTeX |
|---|---|---|---|---|---|---|---|---|
| lower Greek symbol | $\alpha$ | \alpha | $\beta$ | \beta | $\gamma$ | \gamma | $\delta$ | \delta |
| | $\epsilon$ | \epsilon | $\varepsilon$ | \varepsilon | $\zeta$ | \zeta | $\eta$ | \eta |
| | $\theta$ | \theta | $\vartheta$ | \vartheta | $\iota$ | \iota | $\kappa$ | \kappa |
| | $\lambda$ | \lambda | $\mu$ | \mu | $\nu$ | \nu | $\xi$ | \xi |
| | $o$ | o | $\pi$ | \pi | $\varpi$ | \varpi | $\rho$ | \rho |
| | $\iota$ | \iota | $\kappa$ | \kappa | $\varrho$ | \varrho | $\sigma$ | \sigma |
| | $\varsigma$ | \varsigma | $\tau$ | \tau | $\upsilon$ | \upsilon | $\phi$ | \phi |
| | $\varphi$ | \varphi | $\chi$ | \chi | $\psi$ | \psi | $\omega$ | \omega |
| upper Greek symbol | $\Gamma$ | \Gamma | $\Delta$ | \Delta | $\Theta$ | \Theta | $\Lambda$ | \Lambda |
| | $\Xi$ | \Xi | $\Pi$ | \Pi | $\Sigma$ | \Sigma | $\Upsilon$ | \Upsilon |
| | $\Phi$ | \Phi | $\Psi$ | \Psi | $\Omega$ | \Omega | | |
| common math symbol | $\aleph$ | \aleph | $\prime$ | \prime | $\forall$ | \forall | $\exists$ | \exists |
| | $\wp$ | \wp | $\Re$ | \Re | $\Im$ | \Im | $\partial$ | \partial |
| | $\infty$ | \infty | $\nabla$ | \nabla | $\surd$ | \surd | $\angle$ | \angle |
| | $\neg$ | \neg | $\int$ | \int | $\clubsuit$ | \clubsuit | $\diamondsuit$ | \diamondsuit |
| | $\heartsuit$ | \heartsuit | $\spadesuit$ | \spadesuit | | | | |
| binary sign symbol | $\pm$ | \pm | $\cdot$ | \cdot | $\times$ | \times | $\div$ | \div |
| | $\circ$ | \circ | $\bullet$ | \bullet | $\cup$ | \cup | $\cap$ | \cap |
| | $\vee$ | \vee | $\wedge$ | \wedge | $\otimes$ | \otimes | $\oplus$ | \oplus |
| relation math symbol | $\leq$ | \leq | $\geq$ | \geq | $\equiv$ | \equiv | $\sim$ | \sim |
| | $\subset$ | \subset | $\supset$ | \supset | $\approx$ | \approx | $\subseteq$ | \subseteq |
| | $\supseteq$ | \supseteq | $\in$ | \in | $\ni$ | \ni | $\propto$ | \propto |
| | $\mid$ | \mid | $\perp$ | \perp | | | | |
| arrow sign | $\leftarrow$ | \leftarrow | $\uparrow$ | \uparrow | $\Leftarrow$ | \Leftarrow | $\Uparrow$ | \Uparrow |
| | $\rightarrow$ | \rightarrow | $\downarrow$ | \downarrow | $\Rightarrow$ | \Rightarrow | $\Downarrow$ | \Downarrow |
| | $\leftrightarrow$ | \leftrightarrow | $\updownarrow$ | \updownarrow | | | | |

curly brackets should be used. For instance, a^Abc command represents $a^A bc$, where, *A* is the superscript. If Abc are used as superscripts, the command a^{Abc} should be used.

(3) Many LᴬTᴇX expressions are not supported in MATLAB graphics windows. For instance, \frac command is not supported.

LᴬTᴇX scientific typesetting is the most widely used typesetting system in academic world. Interested readers may further refer to [10].

### 6.2.3 Superimposing formulas in plots

There are limitations of using LᴬTᴇX format commands to show mathematics on the plots. For instance, there are a lot of unsupported LᴬTᴇX commands, such as fraction command \frac. Besides, many user-defined LᴬTᴇX commands are not supported and the information such as fonts cannot be assigned the same as in a LᴬTᴇX document. For compatible typesetting tasks, suitable ways are expected to label mathematics on top of the plots.

The overpic package in LᴬTᴇX documentation system can be used to superimpose any LᴬTᴇX command on plots, including mathematics. Any text labels can be placed on top of the plots. This method is recommended to add marks on the plots, rather than using the LᴬTᴇX commands in MATLAB. If the overpic package is used, the label location problems need to be solved, since there are no other tools applicable for this task. A universal MATLAB function is written to solve the problem.

**Example 6.9.** An assistant function for the overpic package is written in MATLAB, for the user to locate and superimpose objects on the plots.

**Solutions.** With the knowledge presented earlier, it is not difficult to write a MATLAB function as follows:

```
function overpic(fname,varargin)
if nargin==0,
    [fname,pathn]=uigetfile('*.eps','Please select the eps file');
    fname=fname(1:end-4);
else, pathn=['D:\xue.dy\BOOKS\MATLAB\Series_in_MATLAB\',...
            'English\Volume1\epsfiles\'];
end
eval([' !epstool -k -q -o' fname '.tif ' pathn  fname '.eps'])
eval(['W=imread(''' fname '.tif'');']); [nh,nw]=size(W);
imshow(~W); figure(gcf), hold off;
eval(['delete([''' fname  '.tif''])' ]), i=0;
while 1, % left click to select, right to terminate
    [x,y,but]=ginput(1); if but~=1, break; end
```

```
      i=i+1; text(x,y,int2str(i))
      nx=fix(x*1000/nw)*0.1; ny=fix((nh-y)*1000/nw)*0.1;
      disp(['\put(' num2str(nx) ',' num2str(ny) '){' int2str(i) '}'])
end
```

The executable `epstool.exe` file is needed, as well as 32-bit Ghostscript software, which is downloadable from

> https://www.ghostscript.com/download/gsdnld.html

The command to use is `overpic`. If no input argument is given, a standard file name dialog box is opened, from which the eps file to be edited can be selected. The file and path names of the eps file to be located are obtained. A new window should be opened, and the figure in the EPS file can be converted into a bitmap image and displayed in the window. The left mouse button can be clicked, and the locating LATEX command can be generated with `\put` command. If the location process is completed, one may click the right mouse button.

If several files in the same folder are to be located, the `pathn` variable in the source code should be modified. Then, function `overpic('file name')` can be used to locate the files.

## 6.3 Other two-dimensional plotting functions

Apart from the standard two-dimensional curves, various other special functions are also provided in MATLAB, as shown in Table 6.3, where the parameters $x$ and $y$ are

**Table 6.3:** Special two-dimensional plotting functions in MATLAB.

| function | meaning | common syntaxes |
|---|---|---|
| bar() | bar plot | bar($x$,$y$) |
| compass() | compass | compass($x$,$y$) |
| feather() | feather plot | feather($x$,$y$) |
| hist() | histogram | hist($y$,$n$) |
| polarplot() | polar plot | polarplot($x$,$y$) |
| stairs() | stair plots | stairs($x$,$y$) |
| semilogx() | $x$-semi-log | semilogx($x$,$y$) |
| comet() | comet plot | comet($x$,$y$) |
| errorbar() | error bar | errorbar($x$,$y$,$y_m$,$y_M$) |
| fill() | filled plot | fill($x$,$y$,c) |
| loglog() | logarithmic | loglog($x$,$y$) |
| quiver() | vector plot | quiver($x$,$y$) |
| stem() | stem plots | stem($x$,$y$) |
| semilogy() | $y$-semi-log | semilogy($x$,$y$) |

respectively the vectors for horizontal and vertical data. Variable c can be used to indicate color options. The vectors $y_m$ and $y_M$ are used for describing the lower and upper bound vectors. Some examples are given to demonstrate the graphics facilities.

### 6.3.1 Polar plots

The definition of the polar coordinate system is given first, then examples are given to show how to draw polar plots using MATLAB.

**Definition 6.1.** The polar coordinate system is a two-dimensional coordinate system in which each point on a plane is determined by a distance $\rho$ from a reference point and an angle $\theta$ from a reference direction. The polar coordinate system is described by an ordered pair $(\rho, \theta)$. A polar equation is usually described by an explicit function $\rho = \rho(\theta)$.

Function `polarplot()` in MATLAB can be used to draw polar plots. If a vector $\theta$ is generated, the $\rho$ vector can be computed from the explicit polar equation, such that `polarplot(θ,ρ)` can be used to draw polar plots, where the unit of $\theta$ is radian. Function `polar()` in the older versions can be used to draw polar plots, however, it is recommended to use the new `polarplot()` function.

**Example 6.10.** Draw a polar plot for the equation $\rho = 5\sin(4\theta/3)$.

**Solutions.** It can be seen that the polar function here is a period function. If the period is not known, an initial vector $\theta$ can be selected, and the vector $\rho$ can be computed. A polar plot can easily be drawn with `polarplot()` as shown in Figure 6.11(a).

```
>> theta=0:0.01:2*pi; rho=5*sin(4*theta/3);
   polarplot(theta,rho) % draw the polar plot
```

It can be seen that the polar plot obtained seems incomplete. This leaves us a question of how to draw a complete polar plot. Many polar equations are periodic; and it can easily be found that the period here is $6\pi$. It one is unable to find the correct period, a vector in a large range can be generated. For instance, $0 \leqslant \theta \leqslant 20\pi$. The complete polar plot can be obtained as shown in Figure 6.11(b).

```
>> theta=0:0.01:20*pi; rho=5*sin(4*theta/3);
   polarplot(theta,rho) % new polar plot
```

**Example 6.11.** Draw the polar plot for an aperiodic function $\rho = e^{-0.1\theta}\sin 3\theta$.

**Solutions.** Selecting the range of $\theta$ as $\theta \in (0, 10\pi)$, the data for the polar plot can be computed and the polar plot can be obtained as shown in Figure 6.12.

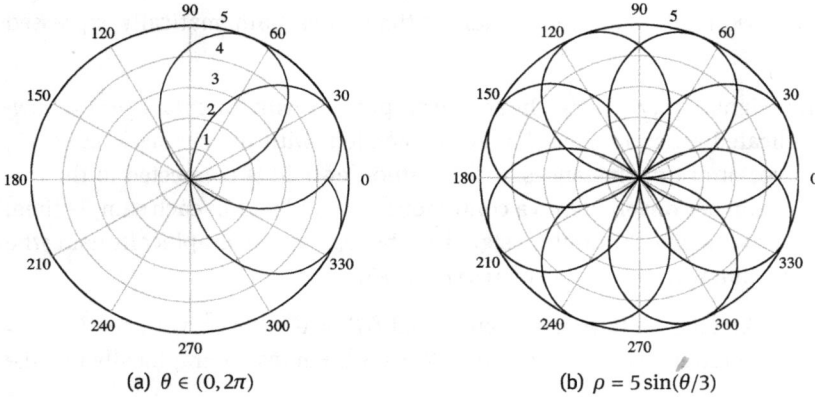

(a) $\theta \in (0, 2\pi)$

(b) $\rho = 5\sin(\theta/3)$

**Figure 6.11:** The polar plots.

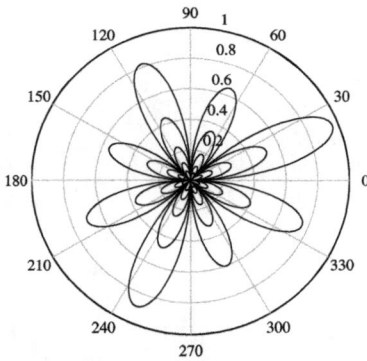

**Figure 6.12:** Polar plot of an aperiodic function.

```
>> theta=0:0.001:10*pi;
   rho=exp(-0.1*theta).*sin(3*theta); polarplot(theta,rho)
```

Since the function is aperiodic, no matter how large the range is selected, the polar plot is incomplete. Luckily, since the value of $\rho$ may vanish for large values of $\theta$ in this example, the current plot depicts the most important information of the polar function.

## 6.3.2 Plots of discrete samples

The definition of a discrete signal is given first, then the representations of the discrete samples are introduced, and also the output signals of zero-order-hold (ZOH) are illustrated.

**Definition 6.2.** Normally, a discrete signal is a time series mathematically expressed as $y_1, y_2, \ldots, y_n$.

A discrete signal can, of course, be shown graphically with function `plot()`, however, the dedicated function `stem()` is recommended, with the syntax `stem(t,y)`, where *t* is a vector of time instances. If a zeroth-order-hold is connected to the discrete signal, it can be converted into a continuous one, and within each sample time, the signal is kept constant. The output signal can be represented graphically using the `stairs()` function, with the syntax of `stairs(t,y)`.

**Example 6.12.** Assume we have a discrete signal $f(t) = e^{-6t} \cos t^2$, with $t = kT$, $k = 0, 1, \ldots, 40$; $T$ is referred to as a sample time, $T = 0.01$. Represent graphically the discrete signal.

**Solutions.** Based on the sample time, a time vector *t* can be generated, and the samples of the discrete function can be evaluated. The discrete stem plot is shown in Figure 6.13.

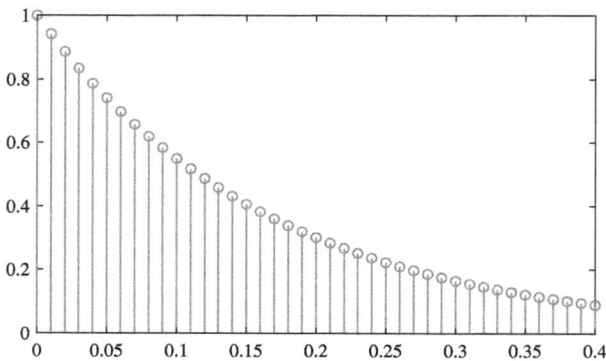

**Figure 6.13:** Stem plot of the discrete samples.

```
>> T=0.01; t=(0:40)*T; f=exp(-6*t).*cos(t.^2); stem(t,f)
```

If function `stairs()` is used to replace the function `stem()`, the stairs plot can be obtained as shown in Figure 6.14.

```
>> stairs(t,y) % draw the stairs plot
```

## 6.3.3 Histograms and pie charts

Histograms and pie charts are important graphical representations in statistics. The definitions of histogram and frequencies are given first, and then examples are given to demonstrate histogram and pie chart representations.

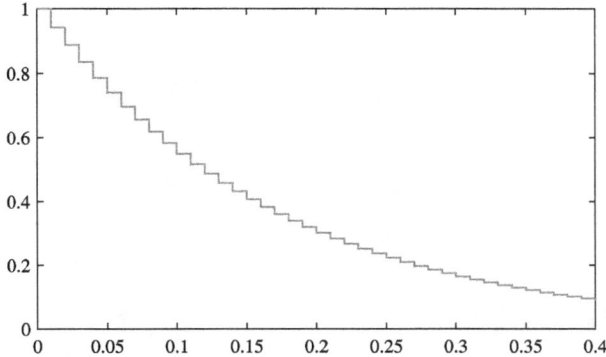

**Figure 6.14:** Stair plot.

**Definition 6.3.** If a set of discrete samples $x_1, x_2, \ldots, x_n$ are generated, and it is known that the data are located in the interval $(a, b)$, divide the interval equally into $m$ subintervals, such that $b_1 = a$, $b_{m+1} = b$. The subinterval is referred to as a bin. Throw each sample $x_i$ into an appropriate bin. Count the number of samples in each bin $(b_j, b_{j+1})$ as $k_j$, $j = 1, 2, \ldots, m$, then $f_j = k_j/n$ is known as the frequency of the bin.

MATLAB function `histogram()` can be used to measure the frequencies of the bins, with the syntaxes

$k$=histogram($\boldsymbol{x}, \boldsymbol{b}$); % the result $k$ is a structured variable
$f$=$k$.Values/$n$; bar($\boldsymbol{b}$(1:end-1)+$\delta$/2,$\boldsymbol{f}$/$\delta$);

where $\delta = x_2 - x_1$ is the width of the equally-spaced bins. Selecting vector $\boldsymbol{b}$, the frequency $\boldsymbol{f}$ can be obtained, from which the histogram can be drawn directly. The length of the frequency vector is one less than that of the $\boldsymbol{b}$ vector. Before using the `bar()` function, the $\boldsymbol{b}$ vector should be moved by half of the width of the bin. In counting the number of sample in each bin, function `histogram()` is recommended, rather than the function `hist()` in the old versions. Examples are used to demonstrate the histograms of the data.

**Example 6.13.** Generate $30\,000 \times 1$ pseudorandom samples for the Rayleigh distribution with $b = 1$. Draw the histogram for the samples and check whether they satisfy Rayleigh distribution or nor.

**Solutions.** Function `raylrnd()` can be used to generate $30\,000 \times 1$ pseudorandom samples. Creating a vector $\boldsymbol{x}$, the function `histogram()` can be used to compute the numbers of the samples falling into each bin. The field `Values` of the structured variable represents the counts. Function `bar()` can then be used to draw the histogram of the distribution, as shown in Figure 6.15. The theoretical probability density function of Rayleigh distribution is also superimposed in the plot. It can be seen that the matching of the two plots is satisfactory.

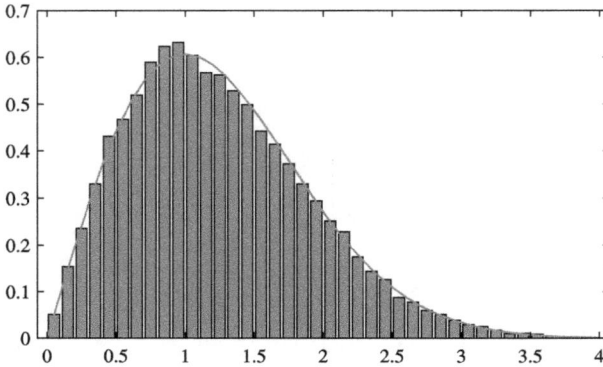

**Figure 6.15:** Probability density function and histogram in Rayleigh distribution.

```
>> b=1; p=raylrnd(1,30000,1); x=0:0.1:4;        % the bins
   y=histogram(p,x); yy=y.Values/(30000*0.1); % histogram data
   x0=x(1:end-1)+0.05; bar(x,yy), y=raylpdf(x,1); line(x,y)
```

The concept of frequency $f$ was shown earlier. It can also be used to draw a pie chart, with the syntax of `pie(`$f$`)`. The percentage of the points in each bin can be depicted with the pie chart.

**Example 6.14.** Consider again the data in Example 6.13. If a pie chart is required, it is not necessary to generate so many bins. The subintervals can be defined as $[0, 0.5]$, $(0, 5, 1], \ldots, (3.5, 4]$. Draw the pie chart in each subinterval.

**Solutions.** The frequency can be computed with the methods shown in the previous example. The pie chart can be drawn directly, in Figure 6.16.

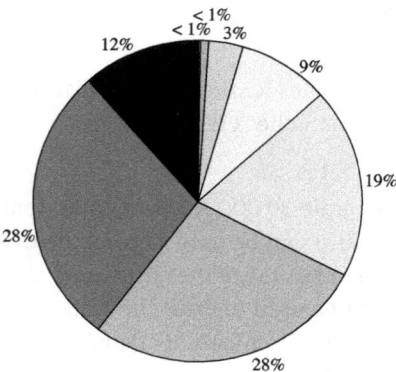

**Figure 6.16:** Pie chart of Rayleigh distribution.

```
>> b=1; p=raylrnd(1,30000,1); x=0:0.5:4;
   y=histogram(p,x); f=y.Values/30000; pie(f), f1=f*100
```

Pie chart representation is straightforward, however, the subinterval information is not displayed in the chart, therefore the frequency information is not presented at all in the chart. The percentage information displayed is also described in a vector $f_1 = [11.5, 28.2, 27.8, 19.1, 8.9, 3.4, 0.8, 0.2] \%$.

### 6.3.4 Filled plots

For a given set of points, $A_1(x_1, y_1)$, $A_2(x_2, y_2)$, ..., $A_n(x_n, y_n)$, a polyline from $A_1$ to $A_2$ to $A_3$ to ... to $A_n$ can be made. An extra segment from point $A_n$ to $A_1$ can be made, such that a closed shape is constructed. With MATLAB function `fill()`, the interior of the region enclosed by a closed-path is filled, so that filled plots can be obtained. This MATLAB function `fill()` is used as `fill(x,y,c)`, where c is the color specification. For instance, 'g' can be used for green; refer to Table 6.1. The argument c can also be represented as a triple, with [1, 0, 0] for red.

If the filled region is defined as the points $A_i$ with $x$ axis, and $x$ vector is sorted in ascending order, two extra points $(x_1, 0)$ and $(x_n, 0)$ can be introduced, such that $x = [x_1, x, x_n]$, $y = [0, y, 0]$. Therefore, the filled plot can be obtained with the `fill()` function.

**Example 6.15.** Consider the Rayleigh distribution studied in Example 6.13, indicate the area which reaches 95 % of area under the probability density function.

**Solutions.** A row vector in $x \in (0, 4)$ can be generated first, and the probability density function of Rayleigh distribution can be evaluated. The key point here is how to find the value of $x$ such that the shaded area is 95 %. With the inverse probability distribution function `raylinv()`, the point $x_0 = 2.4477$ can be found directly. Since at the left end, when $x = 0$, the probability density function is 0, there is no need to have an extra point on the left. While on the right, the point of $x_0$ can be inserted in the vector, and the points whose $x$ value is less than or equal to $x_0$ can be extracted. An point $(x_0, 0)$ can be appended to enclose the expected region. The filled plot is obtained as shown in Figure 6.17.

```
>> x=0:0.1:4; b=1; y=raylpdf(x,b); x0=raylinv(0.95,b)
   ii=x<=x0; x1=[x(ii), x0, x0]; y1=[y(ii),raylpdf(x0,b),0];
   plot(x,y), hold on; fill(x1,y1,'g')
```

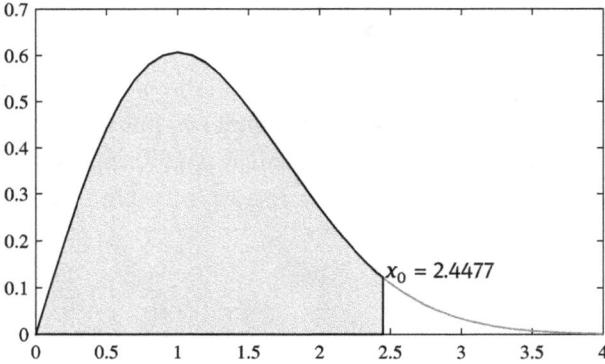

**Figure 6.17:** Rayleigh distribution with 95 % area shaded.

### 6.3.5 Logarithmic plots

In many fields such as in signal processing and automatic control, frequency analysis of systems is always expected, with Bode diagrams commonly used in frequency domain analysis.

**Definition 6.4.** A Bode diagram is the description of the gain $G(s)$ at the frequency points $s = j\omega_1, j\omega_2, \ldots, j\omega_m$, where $\omega_k$'s are referred to as frequency points. The gain $G(j\omega)$ is a complex vector. The relationship of the magnitude $|G(j\omega)|$ and phase $\angle G(j\omega)$, with respect to the frequency vector, can be assessed.

Normally, Bode diagram can be used to describe the magnitude–frequency and phase–frequency plots. Here, for demonstration, only the relationship of magnitude–frequency is considered. The frequency axis is logarithmic, while the vertical axis is the magnitude converted from $20 \lg |G(j\omega)|$, with a unit of decibels (dB).

If the horizontal axis is selected in logarithmic scale, while the vertical one is linear, MATLAB function `semilogx()` can be used to draw the plot. The functions `semilogy()` and `loglog()` can also be used in drawing different logarithmic plots.

**Example 6.16.** Assume that the transfer function of a system is given below. If the frequency range is $\omega \in (0.01, 1\,000)$, drawn the Bode magnitude plot.

$$G(s) = \frac{2(s^{0.4} - 2)^{0.3}}{\sqrt{s}(s^{0.3} + 3)^{0.8}(s^{0.4} - 1)^{0.5}}$$

**Solutions.** Generally, the frequency points should be selected in equally logarithmically scaled format, with function `logspace()`. The frequency response data can be computed directly, logarithmic computation is then carried out to convert the data into decibels. The Bode magnitude plot can be obtained as shown in Figure 6.18.

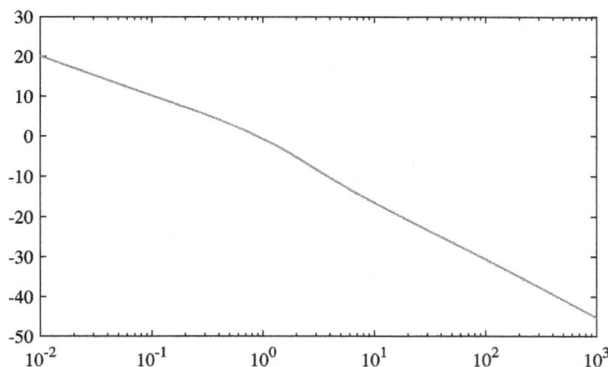

**Figure 6.18:** Bode magnitude plot.

```
>> G=@(s)2*(s.^0.4-2).^0.3./sqrt(s)./...
        (s.^0.3+3).^0.8./(s.^0.4-1).^0.5;
   w=logspace(-2,3,100); M=20*log10(abs(G(1i*w)));
   semilogx(w,M)
```

### 6.3.6 Error bar plots

Error bar plots are widely used in statistical computation. If a set of data $t$ and $y$ is given, and at time $t$, the lower and upper bounds of $y$ are respectively $y_m$ and $y_M$, the error bar plot can be obtained with `errorbar`$(t,y,y_m,y_M)$. If only $y_m$ vector is given, then $y_M$ is automatically set to $-y_m$.

**Example 6.17.** Generate a set of pseudorandom numbers $x$. With function `std()`, the standard deviation can be computed. Draw the error bar plot of the data, if the standard deviation is used as the error bars.

**Solutions.** A set of 15 pseudorandom data can be generated, and the error bar plot can be drawn, as shown in Figure 6.19.

```
>> n=15; x=randn(1,n); t=1:n; e=std(x);
   errorbar(t,x,e*ones(size(x)))
```

### 6.3.7 Dynamic trajectories

The plots discussed so far were static plots. If a curve can be considered as the trajectory of a particle, the result of the particle can be displayed with the plots discussed earlier, while the movement of the particle cannot be observed. If the plotting func-

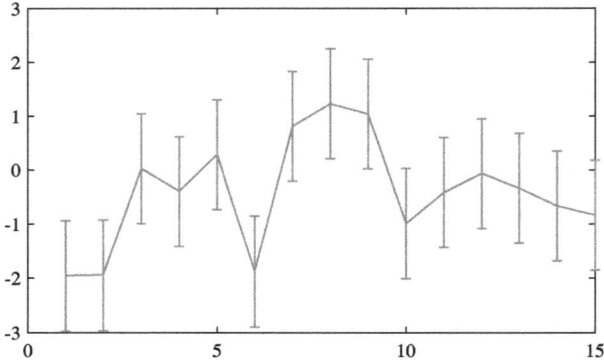

**Figure 6.19:** Error bar plots of the given data.

tion plot() is changed to comet(), a dynamic trajectory display of the particle can be obtained.

**Example 6.18.** Display dynamically the trajectory of the particle in Example 6.1.

**Solutions.** Selecting a step-size of 0.0001, the trajectory of the particle can be computed and displayed dynamically with the following statements:

```
>> x=[-pi:0.0001:pi; y=sin(tan(x))-tan(sin(x)); comet(x,y)
```

### 6.3.8 Two-dimensional animation

With the above illustrated functions, it seems that the plots can be immediately obtained with MATLAB commands. In real applications, it is usually not the case. If a plot statement is followed by a set of intensive computation commands, it can be seen from the current MATLAB mechanism that the plot cannot be immediately drawn, it might be drawn when the intensive commands are executed. It is obvious that this mechanism is not suitable for presenting animation illustrations. A MATLAB function drawnow can be used to suspend the intensive commands, and draw immediately the plot, then the intensive commands can be executed. With such a function, the animation illustrations can be implemented.

Another key point in updating the points in animation is how to make the points in the plot move. If function plot() is used to draw the points, handles can be returned. The data of the objects are stored in the properties 'XData' and 'YData' of the objects. With the function set(), the position information of the objects can be updated, and animation effect can be achieved. The following example can be used to illustrate the animation process of particles.

**Example 6.19.** Consider a swarm of particles in Brownian motion. Assume that the number of particles is $n = 30$, and the interval for observation is $[-30, 30]$. The movement of each particle can be computed from

$$x_{i+1,k} = x_{i,k} + \sigma \Delta x_{i,k}, \quad y_{i+1,k} = y_{i,k} + \sigma \Delta y_{i,k}, \quad k = 1, \dots, n$$

where $\sigma$ is the scaling factor, and the increments $\Delta x_{i,k}$ and $\Delta y_{i,k}$ have standard normal distribution. Simulate Brownian motion with animation method.

**Solutions.** Standard normal distribution pseudorandom data can be generated with randn(), and the scaling factor is selected as $\sigma = 0.3$. Then an infinite loop can be used to animate the particles in vectorized form. To terminate the infinite loop, the Ctrl+C keys can be used.

```
>> n=30; x=randn(1,n); y=randn(1,n); sig=0.3;  % initial points
   figure(gcf), hold off; % bring the existing window to the front
   h=plot(x,y,'o'); axis([-30,30,-30,30])
   while (1), % infinite loop for animation
      x=x+sig*randn(1,n); y=y+sig*randn(1,n);
      set(h,'XData',x,'YData',y), drawnow
   end
```

## 6.4 Plot window partitioning

In real programming, the MATLAB figure window can be divided into several subregions. In each of them, different plots can be drawn. In this section regular partitioning method is introduced first, then arbitrary partitioning is illustrated. Examples are used to demonstrate the partition process and its applications.

### 6.4.1 Regular partitioning

The so-called partitioning means that we divide the whole figure window into $m \times n$ subregions. Regular partitioning method is practical in real applications, where $k$ is the serial number of the partition. An argument can be returned in the function call. With the function subplot(), the figure window can be divided into several subregions, with the syntax subplot$(m,n,k)$, h=subplot$(m,n,k)$, where $h$ is the handle of the partitioned axis.

**Example 6.20.** Draw the Bode diagram of the fractional-order system $G(s)$ in Example 6.16.

**Solutions.** Bode diagram was introduced in Section 6.3.5. A normal Bode diagram is composed of two subplots, the upper one and the bottom one, with subplot() function, i. e., with $2 \times 1$ regions. The upper plot is the Bode magnitude plot, while the lower one is the phase plot. The code in Example 6.16 can be used to draw the magnitude plot, and the new code can be used to draw the phase plot. The results obtained appear in Figure 6.20.

```
>> G=@(s)2*(s.^0.4-2).^0.3./sqrt(s)./...
          (s.^0.3+3).^0.8./(s.^0.4-1).^0.5;
   w=logspace(-2,3,100); subplot(211) % or subplot(2,1,1)
   M=20*log10(abs(G(1i*w))); semilogx(w,M)
   subplot(212), P=angle(G(1i*w))*180/pi; semilogx(w,P)
```

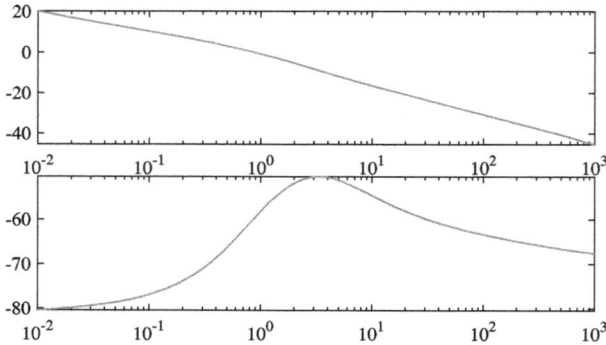

**Figure 6.20:** Bode diagram of fractional-order transfer function.

**Example 6.21.** Draw the sine curve with different functions, in different subregions of the figure window.

**Solutions.** The following statements can be used to draw the plots, as shown in Figure 6.21, where subplot() can be used to divide the figure window into four parts, as a $2 \times 2$ matrix. In each part, a different plot is drawn.

```
>> t=0:.2:2*pi; y=sin(t);           % generate the data first
   subplot(2,2,1), stairs(t,y)      % divide the windows into subregions
   subplot(2,2,2), stem(t,y)        % stem plot
   subplot(2,2,3), bar(t,y)         % bar plot
   subplot(2,2,4), semilogx(t,y)    % horizontal axis in logarithmic scale
```

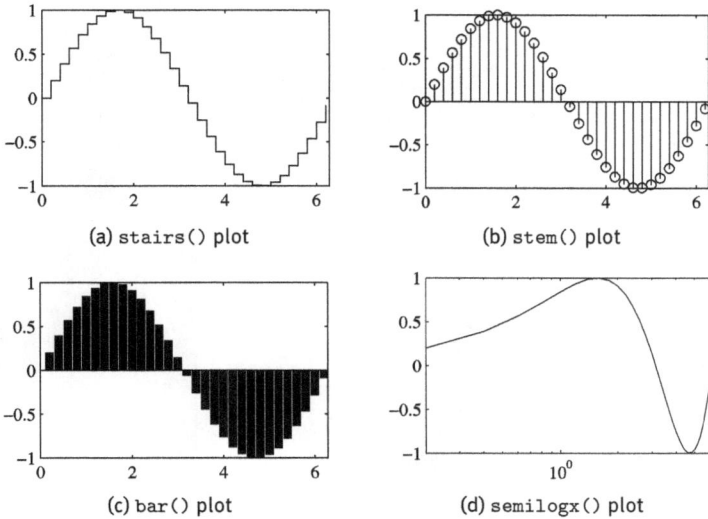

Figure 6.21: Different two-dimensional functions.

## 6.4.2 Arbitrary segmentation

Selecting Insert → Axes menu item, move the cursor to the desired position, and drag the mouse so that an axis can be inserted. The position and size can be adjusted with the mouse. Many axes can be inserted in the same way, and the result is shown in Figure 6.22. Selecting an existing axis, the position and size can be adjusted again with the mouse. Later, window partitioning can be made through the following example.

**Example 6.22.** Draw the curve $f(x) = \sin 1/x$, where $x \in (0,1)$.

**Solutions.** The following statements can be used to draw the curve:

```
>> x=linspace(0,1,1000); y=sin(1./x); plot(x,y)
```

It can be seen that when $x$ is small, the curve is not clear. Therefore, the curves in a smaller interval can be displayed to show details. For instance, the region $x \in [0.02, 0.04]$ can be assigned.

Selecting "Insert → Axes" menu item, a new axis can be established, and the position and size of the axis can be adjusted freely by the user. With the commands given above and zooming facilities, the new results are shown in Figure 6.23. It can be seen that the plot can be enhanced, also the details can be displayed.

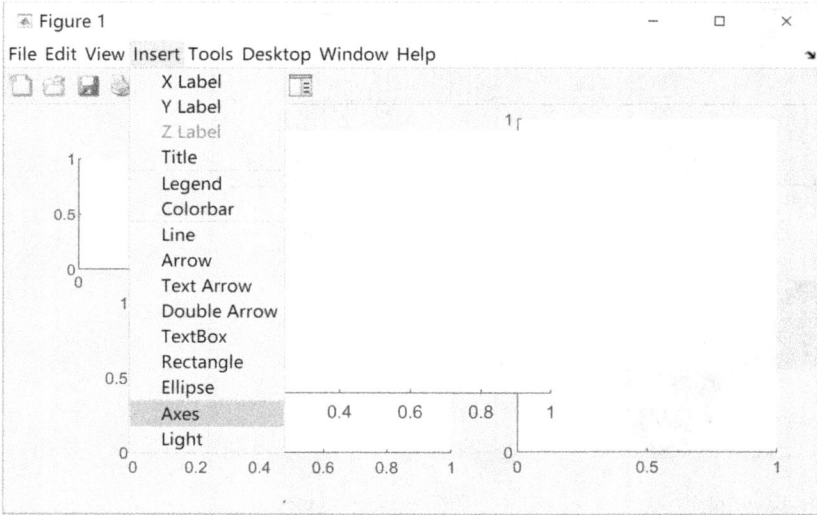

**Figure 6.22:** Insert an axis.

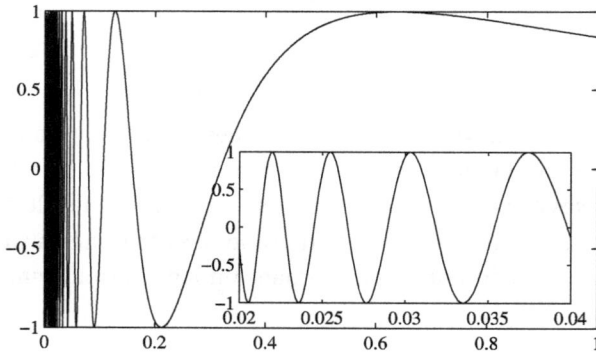

**Figure 6.23:** The axis added with inserted details.

## 6.5 Implicit functions

Implicit functions are mathematically expressed as $f(x,y) = 0$. The curve of such a function cannot be drawn with the methods discussed earlier. For some implicit function, the explicit relationship between variables $x$ and $y$ cannot be obtained. Therefore, we cannot define an $x$ vector, then compute the relevant $y$ vector. Function plot() cannot be used for such implicit functions. The MATLAB function fplot() discussed earlier cannot be used in drawing implicit functions.

MATLAB function `fimplicit()` can be used to draw the curves of the implicit functions, with the syntax of `fimplicit`(implicit expression), where "implicit expression" can be a symbolic expression or an anonymous function. The plotting range can be provided with `fimplicit`(implicit expression, $[x_m, x_M]$), so that the expected curves can be obtained. The default interval is $[-5, 5]$.

In earlier versions of MATLAB, a practical function `ezplot()` could be used in drawing curves of implicit functions, with similar syntax of `fimplicit()`. Strings can be used in `ezplot()` function. It should be noted that `piecewise()` functions cannot be drawn by the `ezplot()` function. Implicit functions can be demonstrated with the following examples.

**Example 6.23.** Draw the curves of the implicit function

$$f(x,y) = x^2 \sin(x + y^2) + y^2 e^{x+y} + 5\cos(x^2 + y) = 0,$$

**Solutions.** It can be seen from the function that the analytical solutions cannot be found and drawn with `plot()` function. With the following statements, the implicit function can be shown in Figure 6.24. It can be seen that plotting an implicit function is simple. The implicit function can be expressed and the plot can be immediately obtained.

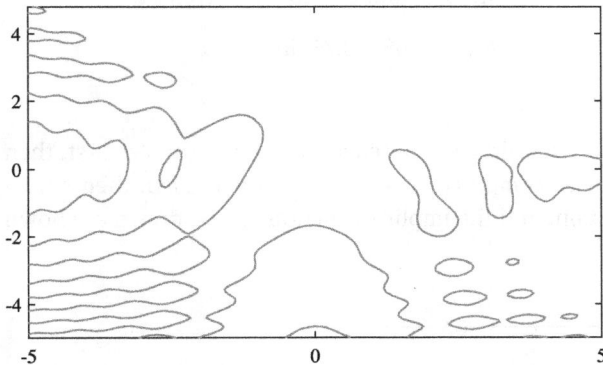

**Figure 6.24:** Implicit function plot.

```
>> syms x y; f=x^2*sin(x+y^2)+y^2*exp(x+y)+5*cos(x^2+y);
   fimplicit(f) % implicit plot
```

The range of $x$ is automatically selected with the function. If one wants to alter the plotting range, the following statements can be used, and the results are obtained as shown in Figure 6.25.

```
>> fimplicit(f,[-10 10]) % the plot in a larger interval
```

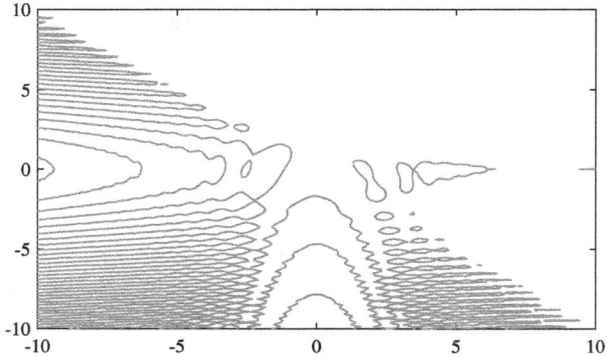

**Figure 6.25:** Implicit function in a larger interval.

An anonymous function can be used to describe an implicit function. Note that the dot operation should be used in vector computation. Function `fimplicit()` can be used, and the result obtained is exactly the same as those obtained earlier.

```
>> f=@(x,y)x.^2.*sin(x+y.^2)+y.^2.*exp(x+y)+5*cos(x.^2+y);
```

**Example 6.24.** Draw the plot of a complicated implicit function given by

$$(r - 3)\sqrt{r} + 0.75 + \sin 8\sqrt{r}\cos 6\theta - 0.75\sin 5\theta = 0$$

where $r = x^2 + y^2$ and $\theta = \arctan(y/|x|)$.

**Solutions.** The variables $x$ and $y$ should be declared as symbolic variables first, then the intermediate variables can be computed. The implicit function can then be defined, with `fimplicit()` function, and the implicit function can be drawn as shown in Figure 6.26.

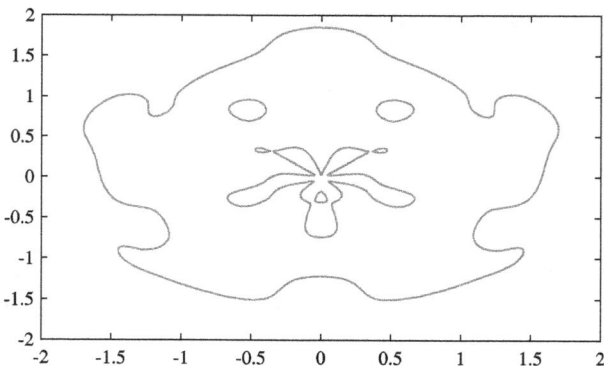

**Figure 6.26:** Plots of a complicated implicit function.

```
>> syms x y; r=x^2+y^2; th=atan(y/abs(x));
   f(x,y)=(r-3)*sqrt(r)+0.75+sin(8*sqrt(r))*cos(6*th)-0.75*sin(5*th);
   fimplicit(f,[-2 2]) % draw the implicit function
```

If an anonymous function is used to describe an implicit function, the intermediate variables $r$ and $\theta$ cannot be computed. The whole implicit function must be described by a single function. Besides, the dot operations must be used in describing the anonymous function. The MATLAB statements are

```
>> f=@(x,y)(x.^2+y.^2-3).*sqrt(x.^2+y.^2)+0.75+...
      sin(8*sqrt(x.^2+y.^2)).*cos(6*atan(y./abs(x)))-...
      0.75*sin(5*atan(y./abs(x)));
   fimplicit(f,[-2 2])
```

## 6.6 Displaying and simple manipulation of images

Bitmap images can be stored in matrices. Each element in the matrix can be regarded as a grayscale pixel. An Image Processing Toolbox in MATLAB can be used to deal with typical problems in image processing. The unsigned 8-bit integers are usually used in storing the grayscale pixels. Color images can typically be described in three primary colors. Three-dimensional arrays $W$ can be used to describe color images, where $W(:,:,1)$ describes the red component in the image, $W(:,:,2)$ and $W(:,:,3)$ store the green and blue components, respectively.

### 6.6.1 Input images

Some image files and file information can be extracted with functions in MATLAB. If one wants to load an image file into MATLAB, the command $W$=imread('file name') can be used. Most mainstream image files can be loaded. The image can be loaded into MATLAB workspace, in variable $W$. If a homochromous image file is loaded, the returned $W$ is a matrix, otherwise, $W$ is a three-dimensional array. The data type is unit8.

Function $F$=imfinfo('file name') can be used to extract basic information of the image file.

**Example 6.25.** Load the color image file tiantan.jpg into MATLAB workspace.

**Solutions.** The following statements can be used to load the color image into MATLAB workspace. With command whos, it can be seen that a three-dimensional array of size $181 \times 444 \times 3$ is loaded, with $181 \times 444$ being the number of pixels in the picture.

```
>> W=imread('tiantan.jpg'); whos W
```

With the command `imfinfo(tiantan.jpg)`, the following image information can be extracted. With the new function `imageinfo()`, similar information can be extracted and displayed as follows:

```
      Filename: 'C:\Users\xuedi\Documents\MATLAB\tiantan.jpg'
   FileModDate: '02-Sep-2008 14:48:40'
      FileSize: 250470
        Format: 'bmp'
 FormatVersion: 'Version 3 (Microsoft Windows 3.x)'
         Width: 444
        Height: 181
      BitDepth: 24
         ...
```

### 6.6.2 Editing and displaying images

Some display and image processing functions are provided in MATLAB. For instance, function `imshow(W)` can be used to directly display the image described in $W$, where $W$ is a matrix, or a three-dimensional array. If function `imtool(W)` is used, an image editing interface can be opened. The images can be displayed and edited in the interface.

**Example 6.26.** Display the image file in Example 6.25.

**Solutions.** With the three-dimensional array $W$, the color image can be displayed with function `imshow()`, as shown in Figure 6.27.

```
>> W=imread('tiantan.jpg'); imshow(W)
```

If function `imtool()` is used to replace `imshow()`, the image editing interface is obtained as shown in Figure 6.28. The user is allowed to display the color information of each pixel.

**Figure 6.27:** Display of tiantan.jpg file.

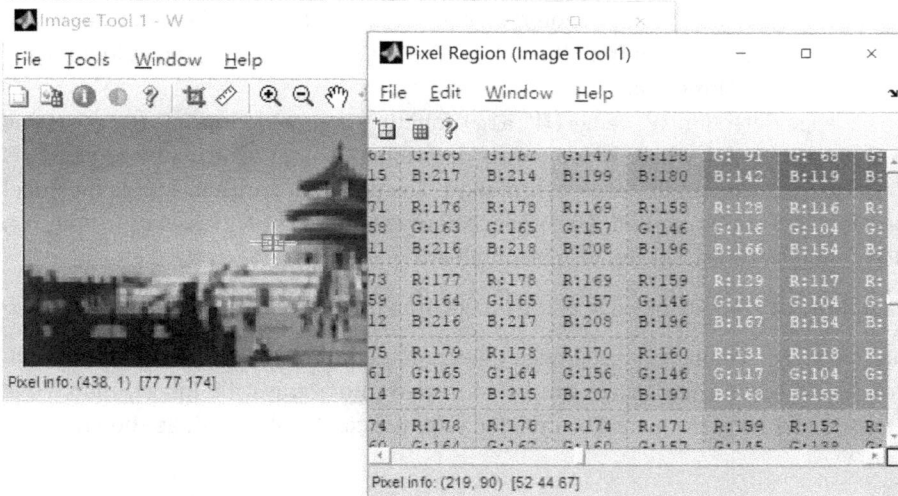

**Figure 6.28:** ImTool editing interface.

With function `imwrite()`, the variable $W$ can be saved into a file.

### 6.6.3 Color space conversion

There are various formats for representing images. The JPG file is the compressed color image in RGB format. Besides, grayscale, binary, color index, and NTSC or YCbCr color images are commonly used. Suitable color space conversions are allowed in MATLAB. For instance, function $W_1$=rgb2gray($W$) can be used to convert an RGB color image into a grayscale one. Function $W_2$=rgb2ind($W$) can be used to convert an RGB color image into a color index one.

**Example 6.27.** Convert the color image in Example 6.25 into a grayscale one.

**Solutions.** With the three-dimensional array $W$, function rgb2gray() can be used to convert a color image into a grayscale one in matrix $W_1$. The matrix $W_1$ is a $181 \times 444$ grayscale one.

```
>> W=imread('tiantan.jpg'); W1=rgb2gray(W); whos W1
```

### 6.6.4 Edge detection

Edge detection is a very important topic in digital image processing and computer vision. If one wants to identify objects from an image, the key step is to extract the edges of the image. The color image should be converted into a grayscale one, from which the edges can be extracted from the image. The edges can be extracted automatically with the MATLAB function.

An edge detection $W_2$=edge($W$,alg) function is provided in Image Processing Toolbox in MATLAB, where $W$ is a matrix describing the grayscale image, while $W_2$ is a binary image containing edge information. The default algorithm, alg, is the Sobel operator, and can be omitted. The other options are 'Prewitt', 'Roberts', 'Canny', and so on. The threshold can also be obtained.

**Example 6.28.** Detect the edge of the image from the one in Example 6.25.

**Solutions.** The commonly used edge detection algorithms are for monochrome images. The original color image can be converted into a grayscale image, then, edge detection can be performed. The inverted edge file can be obtained, as shown in Figure 6.29.

```
>> W=imread('tiantan.jpg'); W1=rgb2gray(W);
   W2=edge(W1); imshow(~W2)
```

**Figure 6.29:** Edge detection results.

### 6.6.5 Histogram equalization

In photography, overexposure and underexposure phenomena often happen. In this case, the histograms of the images may be clustered closely in a condensed interval, and the composition of the image cannot be seen clearly. This phenomenon happens because the histograms may not be well distributed. With the use of histogram equalization methods, the original image can be corrected. Even when the histogram is well distributed, the method can still be used for local processing.

The histogram of the image can be obtained with the imhist() function in the Image Processing Toolbox. If the histogram of the image is unevenly distributed, some part of the image may be clear, while another may be hard to recognize. Sometimes even if the equalized image is better than the original one, the result is not perfect. The adaptive histogram equalization function `adapthisteq()` in the Image Processing Toolbox can be used to find better images.

**Example 6.29.** Consider the picture of the surface of the moon, shown in Figure 6.30(a). The picture is from [9]. The file name is c8fmar.tif. It can be seen that the lower right corner is rather dark, and details cannot be seen clearly. Image processing is needed for this picture. Use histogram equalization method to process the image and observe the results.

**Solutions.** The histogram of the image can be obtained with the Image Processing Toolbox function `imhist()`, and the result can be obtained as shown in Figure 6.30(b). It can be seen that most of the grayscale values are clustered around 0~30. Therefore, the picture on the lower right corner is rather dark, and details of the picture cannot be recognized. Histogram equalization method can be tried to process the image.

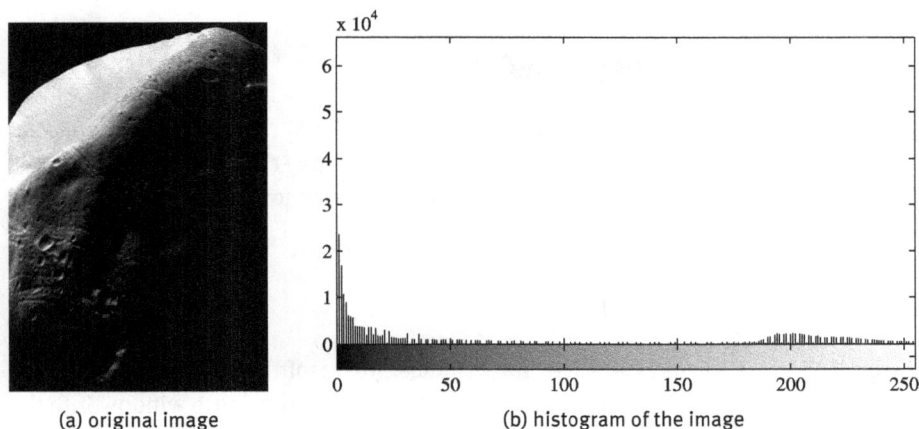

(a) original image  (b) histogram of the image

**Figure 6.30:** Original image and its histogram.

```
>> W=imread('c6fmar.tif'); imhist(W)
```

The adaptive histogram equalization function `adapthisteq()` in the Image Processing Toolbox can be used to find better images. With the following statements, the improved image can be obtained as shown in Figure 6.31(a), and the new histogram is shown in Figure 6.31(b). It can be seen that the improved image is much better and the new histogram is well distributed.

```
>> W1=adapthisteq(W,'Range','original',...
      'Distribution','uniform','clipLimit',0.2);
   imshow(W1); figure; imhist(W1)
```

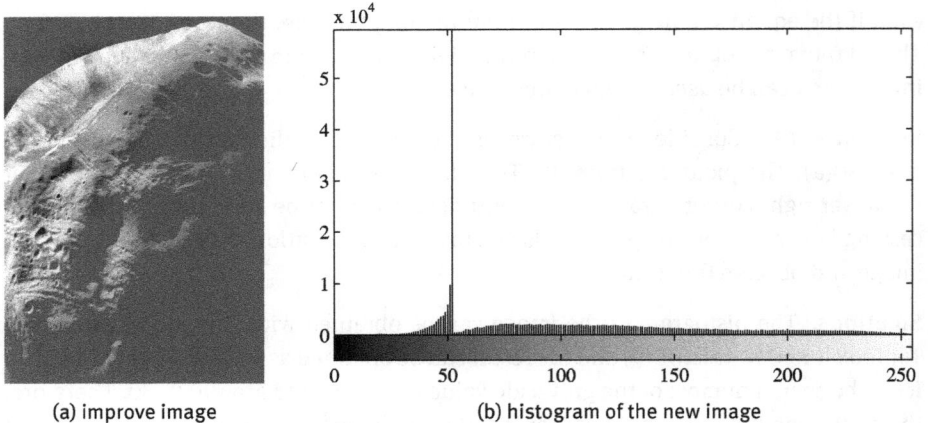

(a) improve image        (b) histogram of the new image

**Figure 6.31:** Image processing with adaptive histogram equalization.

## 6.7 Output of MATLAB graphs

Figures and images can be drawn in a figure window of MATLAB. These figures can be stored, in different format, as figure or image files. They can also be copied to the clipboard. In this section, the output formats of figure windows are presented.

### 6.7.1 Output menus and applications

Since MATLAB is a window-based, for instance, Microsoft Windows, program, the menu item Edit → Copy can be used to copy the figures in the figure window into the clipboard, so that other Windows programs can directly use the figures.

The default copy format is the metafile format. MATLAB also allows copying the figure to the clipboard with the bitmap format. The menu item Edit → Copy Options can be used to set the copy format. Bitmap and metafile formats have their own advantages in representing plots or images. In the metafile format, the drawing information is stored, while in the bitmap format, the pixel information is stored. Generally speaking, the precision in picture drawing is recorded, and the size of the file is usually smaller. No information will be lost when scaling. The sizes of bitmap files are proportional to the number of pixels of the image, and they are not suitable to be scaled. It is recommended to save two-dimensional curves in the metafile format. For three-dimensional graphics, the users may select metafile of bitmap formats according to the size of the file. For images, the bitmap format should be used.

Various figure output options in MATLAB are provided. If the menu item File → Print Preview is selected, a standard dialog box will be shown, allowing the user to set the printer parameters. For instance, the printer type and paper size can be selected.

### 6.7.2 Output commands of plots

Apart from the menu actions in Windows for picture copying, the command `print` can also be used to output figures. For instance, the figure window can be printed directly on a printer. Also, it can be output to files under different format. The command can be called with the syntaxes

`print -ddevice  -options file,   print('-ddevice','-options',file)`

The following options can be used to describe the "devices":
(1) **Direct print.** The option `-dwin` can be used to print the figure on a homochromous printer, while option `-dwinc` can be used to print the figure on a color printer.
(2) **Page description.** The option `-dps` can be used to print the figure as a homochromous PostScript file, `-dpsc` for color PostScript file, `-deps` for homochromous encapsulated PS (EPS) file, and `-depsc` for color EPS files. All the figures in the book are printed as EPS files, which are of better quality than many other formats such as metafiles. Even if Microsoft Word is used, high-quality EPS files can be used instead.
(3) **Common graphical format.** The option `-djpeg` can be used to save the file as a JPEG file, while the option `-dtiff` can be used to save the TIFF file.
(4) **Copy to clipboard.** The option `-dmeta` can be used to copy the plot to the clipboard, with the format of a metafile. If option `-dbitmap` is used, the figure can be copied to the clipboard in the bitmap format.
(5) **Computer setting.** With the option `-dsetup`, a dialog box can be opened, from which the parameters of the computer can be set up.

Many other options can be used in the command `print`. For instance, `-s` can be used to print Simulink models. There are two syntaxes in calling `print` command, one is the use of `print` command, while the other is the use of the `print()` function. The figure can be printed to aaa.eps file. If there exist spaces in the folder names or function names, the function style should be used.

`>> print -deps aaa, print('-deps','aaa')`

## 6.8 Problems

6.1  Draw the curve of the function $y(x) = \sin \pi x/(\pi x)$, where $x \in (-4, 4)$.
6.2  Select a suitable step size to draw the function $\sin 1/t$, where $t \in (-1, 1)$.
6.3  Select a suitable step size to draw the curve of $\tan t$, where $t \in (-\pi, \pi)$. Observe the discontinuities in the function.
6.4  Draw the following functions:
     (1) $f(x) = x \sin x$, $x \in (-50, 50)$,    (2) $f(x) = x \sin 1/x$, $x \in (-1, 1)$.

6.5 For a suitable range of $t$, draw the functions $x = \sin t$, $y = \sin 2t$. Assume that a particle is moving on the curve, draw dynamically the trajectory of the particle.

6.6 In the interval $t \in (0, 2\pi)$, show the three curves, $\sin x$, $\sin 2x$, and $\sin 3x$, in the same coordinates.

6.7 A regular triangle can be drawn by MATLAB statements easily. Use the loop structure to design an M-function so that, in the same coordinates, a sequence of regular triangles can be drawn, each by rotated by a small angle, for instance, $5°$, from the previous one.

6.8 Draw the implicit function $x \sin x + y \sin y = 0$ with $-50 \leqslant x, y \leqslant 50$.

6.9 Draw the following piecewise function:

$$y(t) = \begin{cases} \sin t + \cos t, & t \leqslant 0, \\ \tan t, & t > 0. \end{cases}$$

6.10 For a normal distribution with probability density function

$$p(x) = e^{-(x-\mu)^2/(2\sigma^2)}/(\sqrt{2\pi}\sigma),$$

where $\mu$ is the mean and $\sigma$ is the variance, draw the probability density function for different values of $\mu$ and $\sigma$.

6.11 Generate the first 40 terms from the following sequence:

$$x_k = 1 + \frac{1}{2} + \frac{1}{3} + \cdots + \frac{1}{k} - \ln k,$$

then draw the variation of the sequence with stem() function.

6.12 For the initial values $x_0 = 0$ and $y_0 = 0$, the subsequent terms can be iterated from

$$\begin{cases} x_{k+1} = 1 + y_k - 1.4x_k^2, \\ y_{k+1} = 0.3x_k. \end{cases}$$

If 30 000 iterations are made, compute the vectors $x$ and $y$. The points can be expressed by a dot, rather than line. In this case, the so-called Hénon attractor can be drawn.

6.13 Assume that the power series expansion of a function is

$$f(x) = \lim_{N \to \infty} \sum_{n=1}^{N} (-1)^n \frac{x^{2n}}{(2n)!}.$$

If $N$ is large enough, the power series of $f(x)$ converges to a certain function $\hat{f}(x)$. Please write a MATLAB program that plots the function $\hat{f}(x)$ in the interval $x \in (0, \pi)$. Observe and verify what function $\hat{f}(x)$ is.

6.14 Draw the curve of the function $\sin t^2$ in the interval $t \in (0, \pi)$. If the relationship of the function is not clear in a certain range, zoom the curve for details.

6.15 Select a suitable range of $\theta$, draw the following polar plots:
(1) $\rho = 1.0013\theta^2$, (2) $\rho = \cos 7\theta/2$, (3) $\rho = \sin \theta/\theta$, (4) $\rho = 1 - \cos^3 7\theta$.

6.16 For the parametric equation $x = (1 + \sin 5t/5)\cos t$, $y = (1 + \sin 5t/5)\sin t$, $t \in (0, 2\pi)$, draw the curve. If the value 5 is changed to another value, draw the curve of the parametric equation.

6.17 Find the approximate solutions of the following simultaneous equations:

(1) $\begin{cases} x^2 + y^2 = 3xy^2, \\ x^3 - x^2 = y^2 - y, \end{cases}$
(2) $\begin{cases} e^{-(x+y)^2+\pi/2}\sin(5x + 2y) = 0, \\ (x^2 - y^2 + xy)e^{-x^2-y^2-xy} = 0. \end{cases}$

6.18 Draw the sine function $y = \sin(\omega t + 20°)$, $t \in (0, 2\pi)$, $\omega \in (0.01, 10)$, when $\omega$ changes, also draw the animation of the sine curve.

6.19 Lambert $W$ function is a commonly used function, whose mathematical form is given implicitly as $W(z)e^{W(z)} = z$. Draw the curve of the function.

6.20 Consider the nonlinear difference equation

$$y(t) = \frac{y(t-1)^2 + 1.1y(t-2)}{1 + y(t-1)^2 + 0.2y(t-2) + 0.4y(t-3)} + 0.1u(t).$$

If the input function is a sine function $u(t) = \sin t$, and assuming that the initial value of $y(t)$ is zero and the sample time is $T = 0.05$ seconds, solve the equation and draw the results.

# 7 Three-dimensional graphics

Two-dimensional graphics were covered in the previous chapter. It can be seen that if the data or mathematical model is known, the result can be visualized with graphical methods. It is safe to say that visualization method may enable the users to better understand the data and mathematical expressions.

In this chapter, scientific visualization and application to three-dimensional data and functions are mainly studied. In Section 7.1, the three-dimensional trajectories and curves are presented. Similarly, three-dimensional bar and pie charts, as well as filled and ribbon plots are illustrated. In Section 7.2, three-dimensional surfaces are described. The generation of mesh grid is presented first, followed by three-dimensional mesh and surface plots. Shaded and lighted plots are also presented. Finally, the shading and light models of the surfaces are illustrated. The definition of a viewpoint is presented in Section 7.3, and the setting of a viewpoint is demonstrated to observe the surfaces from any viewpoint. Orthographic views of surfaces are also presented. In Section 7.4, different special three-dimensional plots are presented, including contour lines, quivers, and implicit three-dimensional functions. In Section 7.5, special manipulations of three-dimensional plots are illustrated, where rotation, cutting-off, and image patching are presented. In Section 7.6, four-dimensional graphics are presented, and the MATLAB facilities in volume visualization and animation are discussed.

## 7.1 Three-dimensional curves

The three-dimensional representation of certain functions is usually needed in practical applications, with curves or surfaces, depending totally on the type of mathematical functions and data. In this section, we focus on three-dimensional curves. Also special plots such as filled plots, bar and pie charts, as well as ribbons, are demonstrated.

### 7.1.1 Drawing three-dimensional plots from data

Consider a particle in a three-dimensional space. The spatial position at time $t$ can be represented as $x(t)$, $y(t)$, and $z(t)$. The trajectory of the particle is a three-dimensional plot.

The two-dimensional function `plot()` is extended to a three-dimensional function `plot3()`, which can be used to draw three-dimensional plots, with the syntaxes

```
plot3(x,y,z)
plot3(x₁,y₁,z₁,option 1,x₂,y₂,z₂,option 2,...,xₘ,yₘ,zₘ,option m)
```

https://doi.org/10.1515/9783110666953-007

where "options" are the same as those for two-dimensional plots in Table 6.1. Vectors **x**, **y**, and **z** are the coordinate spatial vectors at time vector **t**.

Similar to the case of two-dimensional curves, three-dimensional graphics facilities can be employed with the functions such as `stem3()` for stem plots, `fill3()` for filled plots, `bar3()` for bar plot, and `comet3()` for dynamical three-dimensional trajectory drawing. The syntaxes of these functions are similar to those in two-dimensional graphics.

**Example 7.1.** The spatial position of a particle is represented by a parametric equation $x(t) = t^3 e^{-t} \sin 3t$, $y(t) = t^3 e^{-t} \cos 3t$, and $z = t^2$. Draw the three-dimensional curve.

**Solutions.** To draw the curve of the parametric equation, a time vector **t** should be generated, and then the three vectors **x**, **y**, and **z** can be computed. With the function `plot3()`, the three-dimensional plot can be obtained as shown in Figure 7.1. Note that the dot operation is used.

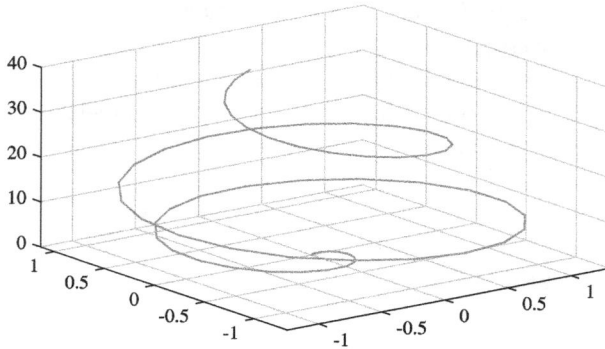

**Figure 7.1:** Three-dimensional plot.

```
>> t=0:.1:2*pi;            % construct a time vector t
   x=t.^3.*exp(-t).*sin(3*t); y=t.^3.*exp(-t).*cos(3*t); z=t.^2;
   plot3(x,y,z), grid   % draw 3D curve and grids
```

If function `stem3()` is used, a three-dimensional stem plot can be superimposed, as shown in Figure 7.2.

```
>> hold on; stem3(x,y,z); % draw stem plot
```

### 7.1.2 Three-dimensional plots of mathematical functions

The curve of the given three-dimensional parametric equations $x(t)$, $y(t)$, and $z(t)$ can also be obtained with function `fplot3()`, with the syntaxes

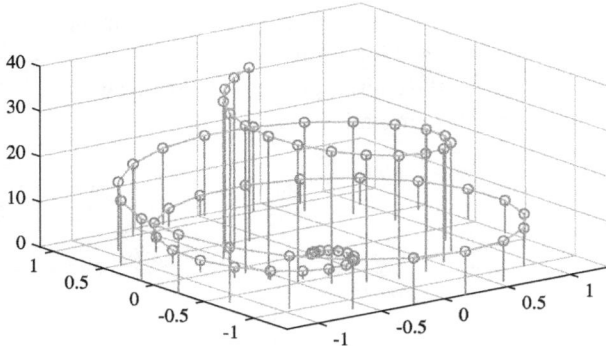

**Figure 7.2:** Three-dimensional plot with stem3() function.

fplot3(xfun,yfun,zfun), fplot3(xfun,yfun,zfun,$[t_m,t_M]$)

where xfun, yfun, and zfun are mathematical expressions, which can be described by symbolic expressions or anonymous functions, with the default interval $[0,5]$ for the independent variable $t$.

**Example 7.2.** Consider again the spatial particle problem in Example 7.1. Draw directly the three-dimensional plot from the mathematical formulas.

**Solutions.** The parametric equations can be described by symbolic expressions, then the plot can be obtained with fplot3(). Identical result to that in Example 7.1 can be obtained.

```
>> syms t; x=t^3*exp(-t)*sin(3*t);
   y=t^3*exp(-t)*cos(3*t); z=t^2; fplot3(x,y,z,[0,2*pi])
```

The parametric equations can also be described by anonymous functions, and the same curve is obtained.

```
>> x=@(t)t.^3.*exp(-t).*sin(3*t); y=@(t)t.^3.*exp(-t).*cos(3*t);
   z=@(t)t.^2; fplot3(x,y,z,[0,2*pi]); fplot3(x,y,z,[0,2*pi])
```

### 7.1.3 Filled plots

If the samples are given in three vectors $x$, $y$, and $z$, the closed polygon can be obtained by joining the first and last sample. The three-dimensional filled plot can be obtained with fill3($x,y,z$,c), where the option c is the same as that in function fill().

**Example 7.3.** Draw the filled plot for the parametric equations in Example 7.1.

**Solutions.** The filled three-dimensional plot can be obtained with the following statements, as shown in Figure 7.3. Green color is used for filling the polygon, however, the filling result is not ideal, since the fillings in the overlapped parts are canceled. In the new software versions, the filling is even more irregular.

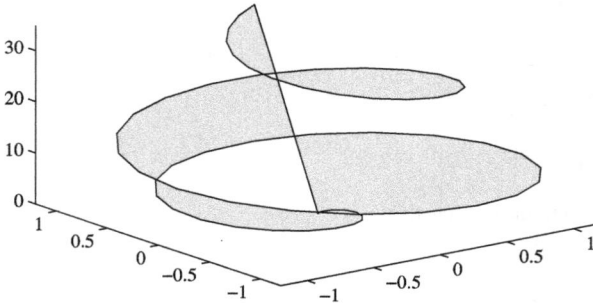

**Figure 7.3:** Three-dimensional filled plot.

```
>> t=0:.1:2*pi;            % construct t
   x=t.^3.*exp(-t).*sin(3*t); y=t.^3.*exp(-t).*cos(3*t); z=t.^2;
   fill3(x,y,z,'g')
```

### 7.1.4 Bar and pie charts

Three-dimensional bar charts are straightforward descriptions of tables. Assume that there is a table whose elements can be represented by matrix $A$. For instance, the rows of $A$ may represent the year, while the columns may represent the product serial, and the elements in the table can be used to represent the quantity of the product. Therefore, function bar3() can be used to represent a three-dimensional bar, with the syntax bar3($A$). The other syntaxes and options of the function are not presented. The users may search for the results with help bar3 command.

A three-dimensional bar chart can be demonstrated with examples.

**Example 7.4.** The production of a company in a three year period is given in Table 7.1. Represent the data with a three-dimensional bar chart.

**Table 7.1:** Production data over a period of three years.

| year | product 1 | product 2 | product 3 | product 4 | product 5 | product 6 |
|------|-----------|-----------|-----------|-----------|-----------|-----------|
| 2015 | 8 500     | 10 500    | 8 000     | 10 000    | 8 500     | 1 100     |
| 2016 | 12 500    | 10 000    | 12 000    | 9 500     | 12 500    | 8 500     |
| 2017 | 8 500     | 12 500    | 9 500     | 11 500    | 9 500     | 10 500    |

**Solutions.** The table can be entered into MATLAB workspace, and the production for each year can be represented in a row, while the production of one product is stored in a column. The following statements can be used to draw a three-dimensional bar chart, as shown in Figure 7.4. It can be seen that the three-dimensional bar chart can be used to represent a matrix in a straightforward way.

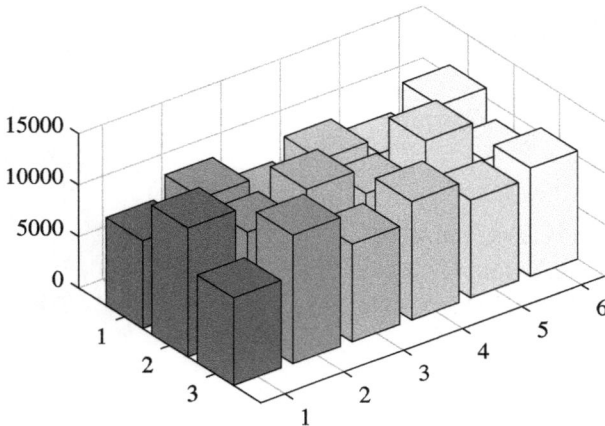

**Figure 7.4:** Three-dimensional bar chart.

```
>> A=[8500,10500,8000,10000,8500,11000;
       12500,10000,12000,9500,12500,8500;
       8500,12500,9500,11500,9500,10500];
   bar3(A)
```

The so-called pie chart here is, in fact, a three-dimensional representation of regular two-dimensional ones. They look much prettier than two-dimensional ones. Special effects in display can be processed. The syntaxes of the function are

pie3($x$), pie3($x$,explode)

where $x$ is a vector. MATLAB can be used to draw charts in a normalized way. The option explode is a binary vector of the same length as that of $x$, with value 0 for normal pies, and 1 for protruding ones. The following example is used to demonstrate three-dimensional pie charts.

**Example 7.5.** Redraw the pie chart in Example 6.14 in the three-dimensional format.

**Solutions.** Pseudorandom numbers can be generated again, and the frequency vector $f$ can be computed, from which the pie chart can be obtained as shown in Figure 7.5. The percentages of the chart are different from those in Example 6.14, since random numbers were used. In the pie chart, the second and fourth sectors are protruding.

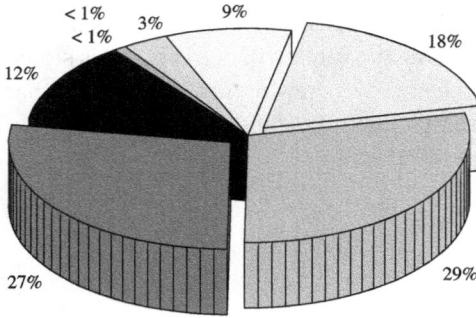

**Figure 7.5:** Three-dimensional pie chart.

```
>> b=1; p=raylrnd(1,30000,1); x=0:0.5:4;
   y=histogram(p,x); f=y.Values/30000; f1=f*100
   key=zeros(size(f)); key([2,4])=1; pie3(f,key)
```

### 7.1.5 Ribbon plots

In real applications, the following problem may be encountered. For a univariate function $y_\alpha = f(x, \alpha)$ with one extra variable $\alpha$, if a fixed value of $\alpha$ is chosen, the function is determined uniquely. Generating $m$ samples of $\alpha$, the curve can be drawn with `plot()` function. Therefore, a group of $m$ plots can be drawn. Since the plots are drawn in the same two-dimensional coordinates, the relationship of the curves with respect to $\alpha$ cannot be distinguished. The plots can be redrawn as three-dimensional.

The function `ribbon()` in MATLAB can be used to complete tasks like this. A vector $x$ can be generated first. For the samples of $\alpha$, a set of column vectors can be computed, and a matrix $Y$ can be established with these columns. With vector $x$ and matrix $Y$, a ribbon plot can be drawn, with the syntaxes

`ribbon(x,Y)`, `ribbon(x,Y,d)`

where $d$ is the width of the ribbons. If the argument is given, the default value of $d = 0.75$ is used. In the ribbon plot obtained, $y$ axis corresponds to vector $x$, while $x$ axis corresponds to the values $1,2,\ldots,m$. It is a pity that the $x$ axis cannot be forced to directly correspond to the actual values of $\alpha$.

**Example 7.6.** For a given sine function $f(t) = \sin t$, its $n$th order derivative can be expressed as $f^{(n)}(t) = \sin(t + n\pi/2)$, where $n$ can also be selected as a non-integer[19]. Use three-dimensional plots to draw the curves of the derivatives for different values of the orders.

**Solutions.** If the orders are selected as 0, 0.1, 0.2, ..., 1, a loop structure can be used to evaluate the fractional-order derivatives. For each $n$, a column vector can be computed for fractional-order derivative. The following statements can be used to draw the ribbon plot, as shown in Figure 7.6.

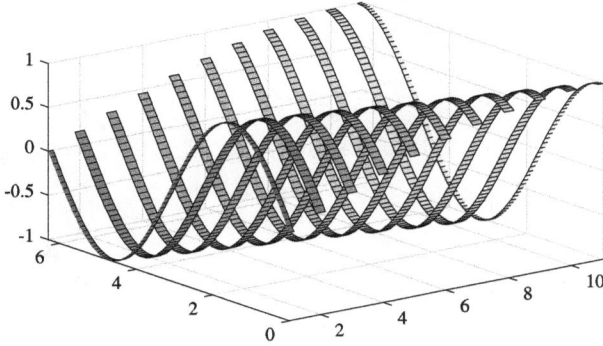

**Figure 7.6:** Ribbon plot of fractional-order derivatives.

```
>> t=linspace(0,2*pi,100)'; Y=[];
   for n=0:0.1:1, Y=[Y, sin(t+n*pi/2)]; end
   ribbon(t,Y,0.3), ylim([0,2*pi]), xlim([1,11])
```

The values for the $x$ axis are marked as 1, 2, ..., 11, where 1 corresponds to $n = 0$, the original function, and 2 stands for $n = 0.1$, the 0.1th order derivative of $f(t)$.

**Example 7.7.** The marked text in the $x$ axis of the plot in Example 7.6 is not informative. How to modify the marks for the correct orders $n$ from the serial numbers?

**Solutions.** This problem can be solved in the editing interface. Clicking the ⃗ button in the toolbar, and selecting View → Property Editor, the edit mode can be launched. Double clicking the axis, the edit window interface can be opened, with the bottom part as shown in Figure 7.7. With the default X Axis, the toolbar can be used.

Clicking the Ticks button, the dialog box is shown in Figure 7.8(a), where the characters in the Labels can be edited freely. For instance, 2 can be modified to 0.1, 4 can

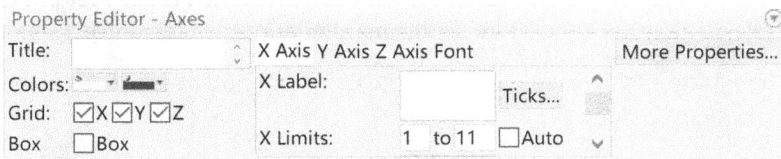

**Figure 7.7:** Editable region of the property inspector.

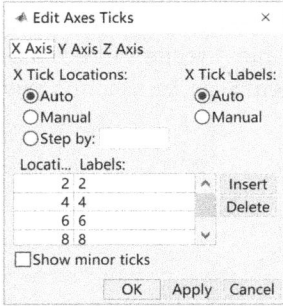

(a) editing the x scales          (b) the edited results

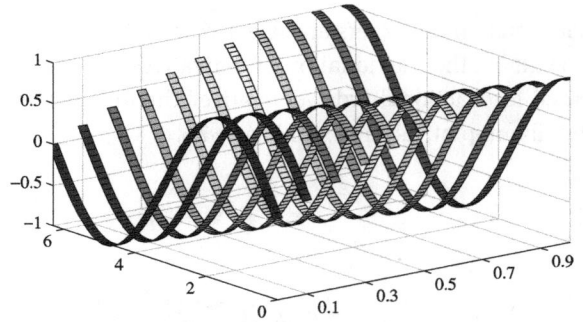

**Figure 7.8:** Conversion of the three-dimensional scales.

be modified to 0.3, and so on. The edited plot obtained is shown in Figure 7.8(b). The characters in Labels can be modified to be any strings.

## 7.2 Three-dimensional surfaces

If a function $z = f(x, y)$ with two independent variables is given, mesh grid points in the $xy$ plane can be generated. The function value at each mesh grid point can be obtained. Based on this information, the surface plot can be obtained.

### 7.2.1 Mesh grids and surfaces

The function `meshgrid()` in Example 4.13 can be used to generate mesh grid points, with two matrices $x$ and $y$ constructed. If the two matrices are stacked up, the coordinates can be represented. The values of the function $z = f(x, y)$ can be computed with the dot operation, yielding a matrix $z$. The three matrices can be used in functions `mesh()` and `surf()`, such that the mesh grid and surface plots can be drawn, with the syntaxes

```
[x,y]=meshgrid(v₁,v₂) % create mesh grids
mesh(x,y,z) or surf(x,y,z) % mesh() mesh grid, surf() surface plot
```

where vectors $v_1$ and $v_2$ are the vectors in the $x$ and $y$ axes. Function `surf()` may also return the handle of the surface, such that further manipulation on it is made possible.

**Example 7.8.** Compute and draw the surface of the fractional-order derivatives of the sine function in Example 7.6.

**Solutions.** Again, the $x$ axis can be used to represent the order $\alpha$, the three-dimensional surface of the result can be obtained directly as shown in Figure 7.9. It can be

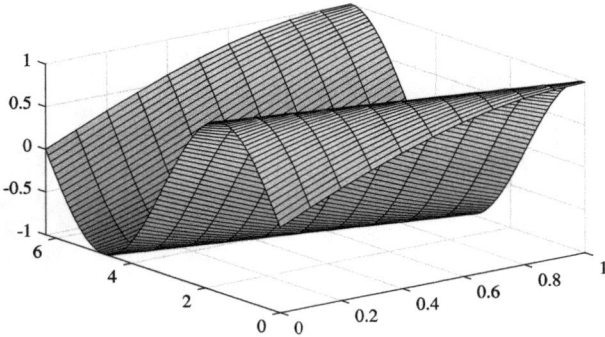

**Figure 7.9:** The surface of the fractional-order derivatives.

seen that, in contrast to the ribbon plot, the labels on the $x$ axis are made informative automatically.

```
>> t=linspace(0,2*pi,100)'; Y=[];
   for n=0:0.1:1, Y=[Y, sin(t+n*pi/2)]; end
   surf(0:0.1:1,t,Y), ylim([0,2*pi]), xlim([0,1])
```

**Example 7.9.** For the given function $z = f(x,y) = (x^2 - 2x)e^{-x^2-y^2-xy}$, select a region in the $xy$ plane, then draw the surface of the function.

**Solutions.** Function `meshgrid()` can again be used to generate mesh grids in the $xy$ plane. The $x$ axis is assigned in the interval $(-3, 2)$, with an increment of 0.1, and the $y$ axis is in the interval $(-2, 2)$, with an increment of 0.1. The value of the function can be computed in the matrix $z$. Function `mesh()` can be used to draw the mesh grid plot, as shown in Figure 7.10.

```
>> [x,y]=meshgrid(-3:0.1:2,-2:0.1:2);   % mesh grid matrices x and y
   z=(x.^2-2*x).*exp(-x.^2-y.^2-x.*y);  % compute the matrix z
   mesh(x,y,z)                          % draw mesh grid plot
```

If the function `mesh()` is substituted by function `surf()`, the surface plot can be obtained as shown in Figure 7.10. It can be seen that each of the mesh grids is colored automatically, as shown in Figure 7.11.

```
>> surf(x,y,z)    % draw the three-dimensional surface
```

**Example 7.10.** Draw the surface plot of the following function:

$$z = f(x,y) = \frac{1}{\sqrt{(1-x)^2 + y^2}} + \frac{1}{\sqrt{(1+x)^2 + y^2}}.$$

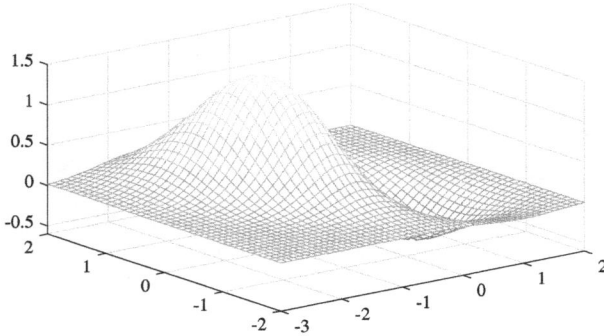

**Figure 7.10:** Mesh grid plot of the function.

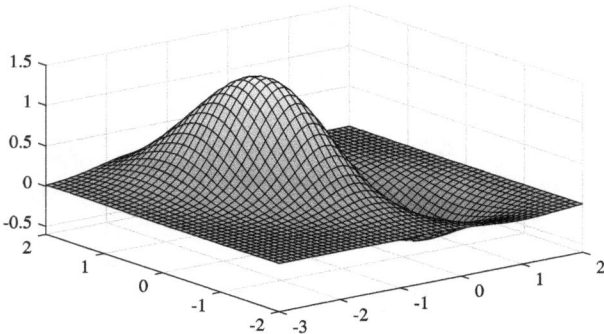

**Figure 7.11:** Surface plot with the function `surf()`.

**Solutions.** The following statements can be used to draw the three-dimensional surface, as shown in Figure 7.12.

```
>> [x,y]=meshgrid(-2:.1:2);   % generate the mesh grid
   z=1./(sqrt((1-x).^2+y.^2))+1./(sqrt((1+x).^2+y.^2));
   surf(x,y,z),   % draw surface plot
```

In fact, the surface obtained has some problems. At points $(\pm 1, 0)$, the value of the function is $\infty$. The increment such regions should be reduced, with variable step-size method. The surface of the plot can be obtained as shown in Figure 7.13. The same ranges of the axes can be obtained as in Figure 7.12. Note that the function values at $(\pm 1, 0)$ are infinite.

```
>> xx=[-2:.1:-1.2, -1.1:0.02:-0.9, -0.8:0.1:0.8, ...
       0.9:0.02:1.1, 1.2:0.1:2];
   yy=[-1:0.1:-0.2, -0.1:0.02:0.1, 0.2:.1:1];
```

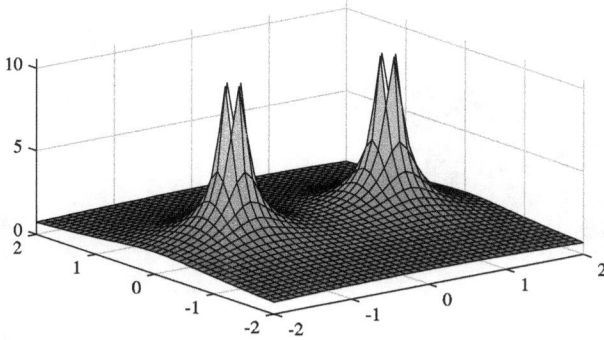

**Figure 7.12:** Three-dimensional surface.

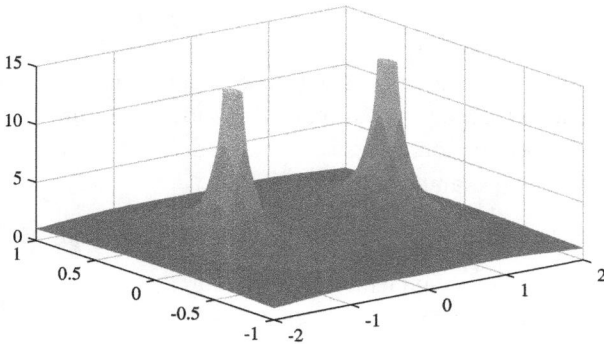

**Figure 7.13:** Surface plot with variable step-size mesh grids.

```
[x,y]=meshgrid(xx,yy); % generate grids
z=1./(sqrt((1-x).^2+y.^2))+1./(sqrt((1+x).^2+y.^2));
surf(x,y,z), shading flat; zlim([0,15]) % 3D surface
```

**Example 7.11.** Draw the surface of the function $f(x,y) = x \sin 1/y + y \sin 1/x$ in the interval $-0.1 \leqslant x, y \leqslant 0.1$.

**Solutions.** The mesh grid points can be generated with $[x,y]$=meshgrid(-0.1: 0.002:0.1). However, the generated mesh grid includes the point $x = 0$ and $y = 0$, and this may lead to troubles. A tiny offset can be introduced to avoid the two lines $x = 0$ and $y = 0$. The offset introduced is $\delta = 10^{-5}$. The following statements can be used to draw the surface, as shown in Figure 7.14.

```
>> [x y]=meshgrid((-0.1+1e-5):0.002:0.1);
   z=x.*sin(1./y)+y.*sin(1./x); surf(x,y,z)
```

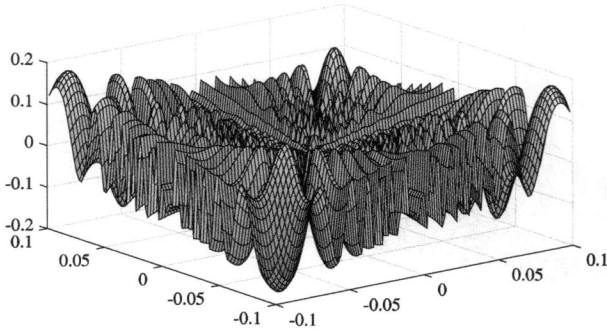

**Figure 7.14:** Three-dimensional surface with an offset introduced.

## 7.2.2 Shading and lights

MATLAB command `shading` can be used to draw a shaded three-dimensional surface. Using the option `flat`, with no grid lines, the result is obtained as shown in Figure 7.15. The option `interp`, with smooth interpolation, can also be applied. The shaded surface can be obtained with the following examples.

**Example 7.12.** Using the option `flat`, draw again the shaded surface in Example 7.9.

**Solutions.** The surface can be drawn again, and with `shading` command, the shaded surface can be obtained as shown in Figure 7.15. If the option `flat` is changed to `interp`, observe the new shading result.

```
>> [x,y]=meshgrid(-3:0.1:2,-2:0.1:2);    % mesh grid matrix x and y
   z=(x.^2-2*x).*exp(-x.^2-y.^2-x.*y);   % compute the height z
   surf(x,y,z), shading flat
```

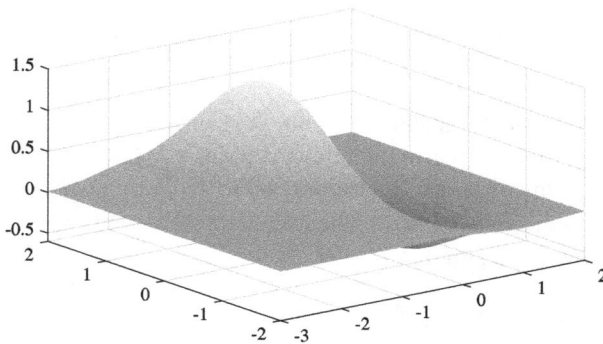

**Figure 7.15:** The shaded surface with option `shading flat`.

The following functions in MATLAB are employed to display surfaces with different effects. For instance, function `light()` can be used to show lighting results. A light source can be placed at $(x,y,z)$, with the command `light('Position',[x,y,z])`. Function `material` can be used to modify the surface material, for instance, the function `material metal` can be used to define a metal surface, while `material shiny` can be used to defined reflective materials. The `help` command can be used to show the commands and functions, and the surfaces with different options can be used to show the decorating effects.

**Example 7.13.** Adding a light source at $(3,2,20)$, draw the surface in Example 7.9 with metal material.

**Solutions.** A light source can be added with the following statements. Also the material of the surface can be assigned, and the new surface is shown in Figure 7.16. Different light source positions and materials can be assigned, and the readers may observe the results in different combinations.

```
>> [x,y]=meshgrid(-3:0.1:2,-2:0.1:2);    % generate mesh grids
   z=(x.^2-2*x).*exp(-x.^2-y.^2-x.*y);   % compute the function
   surf(x,y,z), shading flat,
   light('Position',[3,2,20]), material metal
```

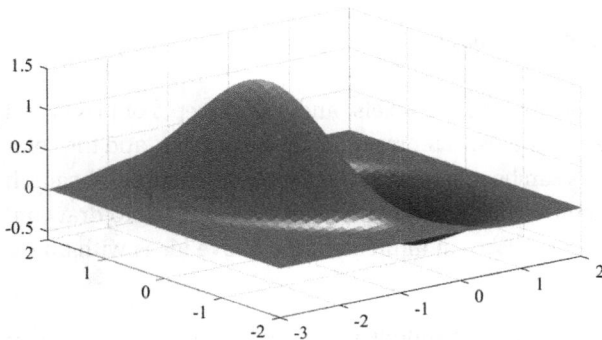

**Figure 7.16:** Surface with light added.

Other functions such as `surfc()` and `surfl()` can be used, where apart from the surface, contours or light effects can also be drawn. Function `waterfall()` can be used to draw the surface with a waterfall effect. Contour lines can also be drawn with functions such as `contour()` and `contour3()`. Examples are introduced here for demonstration.

**Example 7.14.** Use function `waterfall()` to display the surface in Example 7.9.

**Solutions.** If the plotting function is substituted by `waterfall()`, the result obtained is shown in Figure 7.17. Of course, one may try different commands to represent the data.

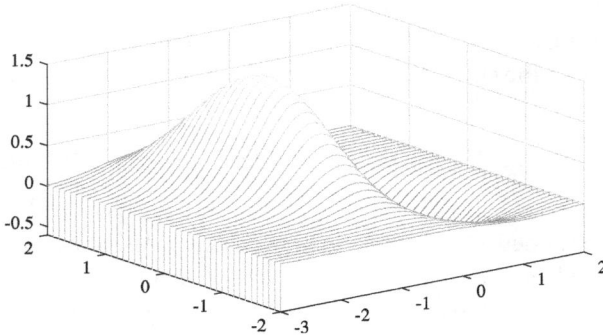

**Figure 7.17:** Surface with `waterfall()` function.

```
>> [x,y]=meshgrid(-3:0.1:2,-2:0.1:2);   % mesh grids
   z=(x.^2-2*x).*exp(-x.^2-y.^2-x.*y);   % function values
   waterfall(x,y,z)
```

### 7.2.3 Three-dimensional surface from images

It has been indicated that images are made of pixels, and the concepts of pixels and mesh grids are similar. The area of interest is represented by mesh grids, and the color or grayscale values in the images can be regarded as the functions values of the mesh grids. Therefore the surface facilities presented earlier can also be used to draw images, with functions such as `surf()`. We can implement the above ideas with an example.

**Example 7.15.** Consider the image used in Example 6.25. Use function `surf()` to display the image, as the surface of a function.

**Solutions.** It has been pointed out that if color images are loaded into MATLAB workspace, three-dimensional arrays can be obtained. If it is converted into a grayscale image, a $181 \times 444$ matrix can be found, with the data type of `uint8`. Function `surf()` cannot be used for this data type. It should be converted into a double precision data type, but before that the grayscale data should be converted and subtracted from 255. The following commands can be used to draw the surface display of the grayscale image as shown in Figure 7.18. It can be seen that the embossment finish can be obtained.

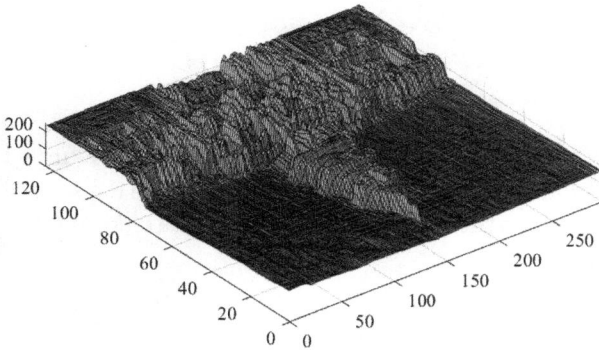

**Figure 7.18:** Embossment display of images (after rotation).

```
>> W=imread('tiantan.jpg'); W1=rgb2gray(W); W0=255-W1;
   surf(double(W0)), ylim([0 181]), xlim([0 444])
```

### 7.2.4 Representation of functions

If the explicit expression $z = f(x,y)$ is known, anonymous functions and symbolic expressions can both be used to describe the function. Then function `fsurf()` can be used to draw the surface of the function, with the syntaxes

`fsurf(f)`, `fsurf(f,[xₘ,xₘ])`, `fsurf(f,[xₘ,xₘ,yₘ,yₘ])`

fsurf$(f)$, fsurf$(f,[x_m,x_M])$, fsurf$(f,[x_m,x_M,y_m,y_M])$

where $f$ is the mathematical expression, and the range of the plots in the $x$ and $y$ axes can be assigned. If none of them are specified, the default range is $[-5,5]$.

**Example 7.16.** Assume that the joint probability density function is described as the following piecewise function[2]:

$$p(x_1,x_2) = \begin{cases} 0.5457\exp(-0.75x_2^2 - 3.75x_1^2 - 1.5x_1), & x_1 + x_2 > 1, \\ 0.7575\exp(-x_2^2 - 6x_1^2), & -1 < x_1 + x_2 \leqslant 1, \\ 0.5457\exp(-0.75x_2^2 - 3.75x_1^2 + 1.5x_1), & x_1 + x_2 \leqslant -1. \end{cases}$$

Display the function with a three-dimensional surface.

**Solutions.** Selecting $x = x_1$ and $y = x_2$, the function values can be generated with loop and conditional structures. However, this is not a good way to represent piecewise functions. Recalling the method to describe piecewise functions discussed earlier, relationship expressions can be used as well for two-dimensional problems.

```
>> [x,y]=meshgrid(-1:.04:1,-2:.04:2); % mesh grids
   z= 0.5457*exp(-0.75*y.^2-3.75*x.^2-1.5*x).*(x+y>1)+...
```

```
       0.7575*exp(-y.^2-6*x.^2).*((x+y>-1) & (x+y<=1))+...
       0.5457*exp(-0.75*y.^2-3.75*x.^2+1.5*x).*(x+y<=-1);
 h=surf(x,y,z), shading flat   % surface with handle h
```

The surface plot in Figure 7.19 can be obtained. Besides, since a handle h is returned in the surf() function call, command delete(h) can be used to remove the surface directly. Also manipulations on objects such as surface rotation will be studied later.

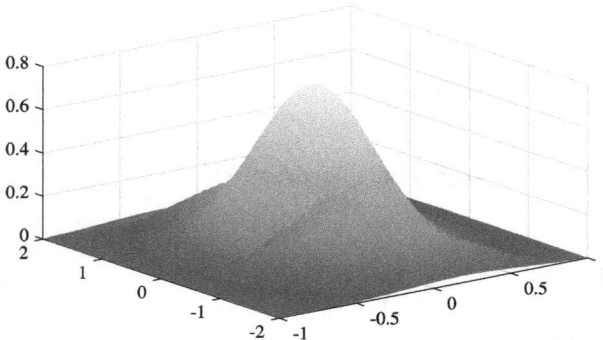

**Figure 7.19:** Surface of a two-dimensional piecewise function.

If an anonymous function or symbolic expression is used to describe the piecewise function, function fsurf() can be used to draw its surface, and the result obtained is exactly the same as that obtained earlier.

```
>> f=@(x,y)0.5457*exp(-0.75*y.^2-3.75*x.^2-1.5*x).*(x+y>1)+...
       0.7575*exp(-y.^2-6*x.^2).*((x+y>-1) & (x+y<=1))+...
       0.5457*exp(-0.75*y.^2-3.75*x.^2+1.5*x).*(x+y<=-1);
   h=fsurf(f)
```

Many simple and concise functions are provided in MATLAB. For instance, the function fmesh(), and also those in earlier versions such as ezsurf() and ezmesh(). If an explicit mathematical function is given, these functions can be used directly to find the expected three-dimensional plots.

### 7.2.5 Surfaces from scattered data

In actual scientific research, usually the mesh grid data may not be available, or may be too costly to obtain. Interpolation technique can be adopted to reconstruct mesh grid data so as to draw the surfaces from the given data.

If the scattered three-dimensional data are known and stored in the vectors $x$, $y$, and $z$, function `meshgrid()` can be used to generate mesh grid interpolation points in matrices $x_1$ and $y_1$. Function $z_1$=`griddata`$(x,y,z,x_1,y_1,$'v4'$)$ can then be used to acquire mesh grid data. Function `surf`$(x_1,y_1,z_1)$ can be used then to draw the surfaces.

**Example 7.17.** Generate several random samples for the function in Example 7.9, and then draw the surface of the function from the samples.

**Solutions.** Let 100 random samples be generated and the values of $z$ computed. The distribution of the samples in the $xy$ plane is shown in Figure 7.20. It can be seen that the samples are distributed relatively uniformly.

```
>> n=100; x=-3+5*rand(n,1); y=-2+4*rand(n,1); % random samples
   z=(x.^2-2*x).*exp(-x.^2-y.^2-x.*y); plot(x,y,'o')
```

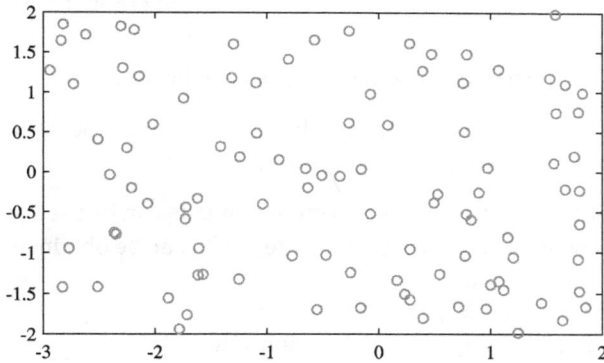

**Figure 7.20:** The distribution of the samples in the $xy$ plane.

Interpolation from the scattered samples can be used so as to reconstruct the surface of the function. The results obtained are exactly the same as those obtained before, or the differences cannot be witnessed.

```
>> [x1,y1]=meshgrid(-3:0.2:2,-2:0.2:2);
   z1=griddata(x,y,z,x1,y1,'v4'); surf(x1,y1,z1)
```

## 7.3 Viewpoint setting in three-dimensional plots

Viewpoints can be modified by the facilities in MATLAB, where an arbitrary viewpoint can be assigned by the user. There are two ways in assigning the viewpoints. One is with the toolbar buttons discussed earlier to change visually the viewpoints. An alternative way is to use function view() to set the viewpoint. In this section, the definition of the viewpoint is presented first, followed by the methods of arbitrary viewpoint settings.

### 7.3.1 Definition of viewpoints

The definitions of the viewpoint angles are defined in MATLAB as shown in Figure 7.21(a). Viewpoints can be uniquely determined with two angles, azimuth and elevation, defined below. The current angles can be extracted from MATLAB command $[\alpha,\beta]$=view(3).

**Definition 7.1.** Draw a line between the viewpoint and the origin. The azimuth angle $\alpha$ is defined as the angle between the projection of the line on the $xy$ plane, and the negative direction of the $y$ axis. The default value is $\alpha = -37.5°$. The elevation angle $\beta$ is defined as the angle between the line and its projection on the $xy$ plane, with a default value of $\beta = 30°$.

If the viewpoint is to be changed, the command view$(\alpha,\beta)$ can be used.

**Example 7.18.** For the surface in Figure 7.19, if the azimuth angle is set to $\alpha = 80°$, and elevation is set to $\beta = 10°$, draw the surface again with the new viewpoint.

**Solutions.** For the above mentioned angles, the following statements can be used to draw the surface and set the viewpoint. The surface in Figure 7.21(b) can be obtained.

```
>> [x,y]=meshgrid(-3:0.1:2,-2:0.1:2);        % mesh grids
   z=(x.^2-2*x).*exp(-x.^2-y.^2-x.*y);       % function values
   surf(x,y,z), view(10,80), xlim([-3,2])    % viewpoint
```

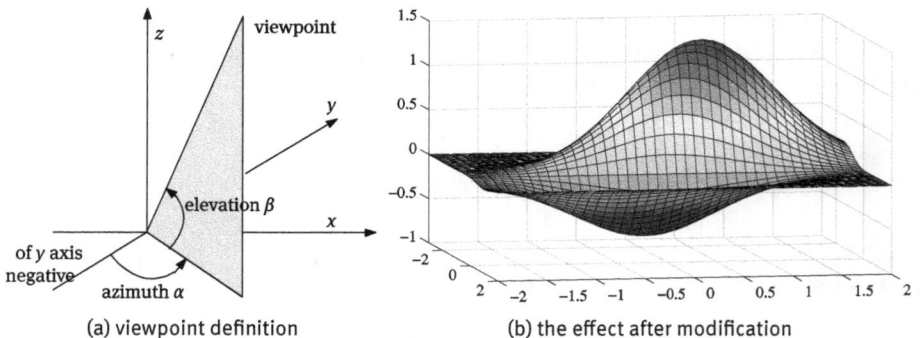

(a) viewpoint definition      (b) the effect after modification

**Figure 7.21:** Viewpoint definition and modification.

## 7.3.2 Orthographic views

The so-called orthographic views of the three-dimensional plots are the planform, side elevation, and front view. These concepts are important in descriptive geometry and computer graphics. With three-dimensional surfaces, the orthographic views are usually expected to better display the three-dimensional objects.

The planform view is the top–down viewpoint for the object. The elevation angle $\beta$ is set to 90°, and there is no need to adjust the azimuth angle $\alpha$, such that it can be set to 0°. Therefore when a three-dimensional surface is displayed, the command view(0,90) in MATLAB can be used. Similarly, if a front view is expected, the command view(0,0) should be used, while if side elevation is needed, the command view(90,0) can be used. If one wants to view the surface from the left, the command view(-90,0) can be used.

**Example 7.19.** Draw the orthographic views for the surface in Example 7.9.

**Solutions.** With the following statements, the three-dimensional plots can be drawn, and different viewpoints can be set as shown in Figure 7.22.

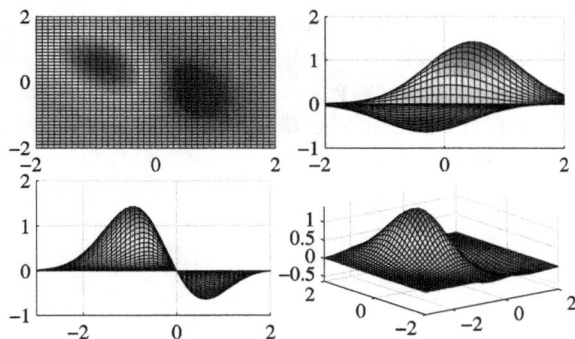

**Figure 7.22:** The surface with orthographic views.

```
>> [x,y]=meshgrid(-3:0.1:2,-2:0.1:2);
   z=(x.^2-2*x).*exp(-x.^2-y.^2-x.*y); subplot(224), surf(x,y,z)
   subplot(221), surf(x,y,z), view(0,90);   % planform
   subplot(222), surf(x,y,z), view(-90,0);  % side elevation
   subplot(223), surf(x,y,z), view(0,0);    % front view
```

## 7.3.3 Setting of arbitrary viewpoints

A button ◎ is provided in the toolbar of the graphics window. If the button is clicked, dragging the mouse, the viewpoints can be redefined, such that ideal observation of three-dimensional plots can be made. In the viewpoint setting process, the two key angles are also displayed on the lower-left corner of the graphics window. Also function $[\alpha,\beta]$=view(3) can be used to define the viewpoint.

## 7.4 Other three-dimensional plots

Apart from the traditional curves and surfaces for three-dimensional plots, plots such as contours, quiver plots, Voronoi diagrams, Delaunay triangulation, spheres, and cylinders are often expected. Besides, the surfaces of three-dimensional implicit functions are also sometimes required.

### 7.4.1 Contour lines

Contour lines are also known as isopleths in geography, referring to closed-paths of the points with the same heights. If the three-dimensional surface is regarded as the terrain, the contour lines can be defined.

Assuming that the mesh grid data of a three-dimensional surface are specified in matrices $x$, $y$, and $z$, function contour() can be used to draw the contours from the three-dimensional data, with the syntax contour($x,y,z,n$), where $n$ is the number of contour lines. An alternative syntax of the function is $[C,h]$=contour($x,y,z,n$), the returned argument h is the contour information, while $C$ returns the contour matrix. With this information, function clabel($C$,h) can be used to superimpose height information on the contour lines.

Function contourf() can be used to generate filled contours, while function contour3() can be used to draw three-dimensional contours, with respectively the following syntaxes:

contourf($x,y,z,n$) and contour3($x,y,z,n$)

**Example 7.20.** Considering the piecewise function in Example 4.20, draw the contours for the given function.

**Solutions.** The following statements can be used and the contours can be drawn directly with function contour(), as shown in Figure 7.23.

```
>> [x,y]=meshgrid(-1:.1:1,-2:.1:2);   % generated mesh grids
   z= 0.5457*exp(-0.75*y.^2-3.75*x.^2-1.5*x).*(x+y>1)+...
      0.7575*exp(-y.^2-6*x.^2).*((x+y>-1) & (x+y<=1))+...
      0.5457*exp(-0.75*y.^2-3.75*x.^2+1.5*x).*(x+y<=-1); % piecewise
   contour(x,y,z);
```

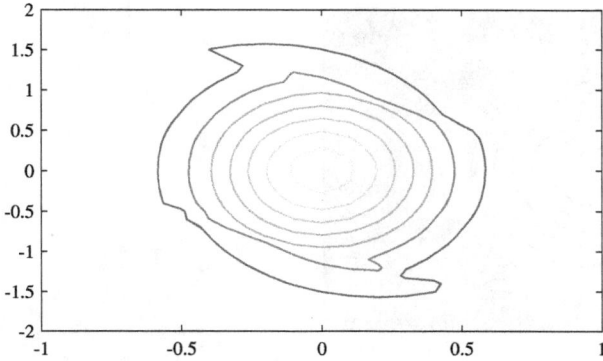

**Figure 7.23:** Contours of a piecewise function.

Apart from contours themselves, the information of the contours is returned as contour information h, and also in the contour matrix **C**. Based on the two arguments, the heights can be superimposed on top of the plot, as shown in Figure 7.24.

```
>> [C,h]=contour(x,y,z); clabel(C,h) % contours with heights
```

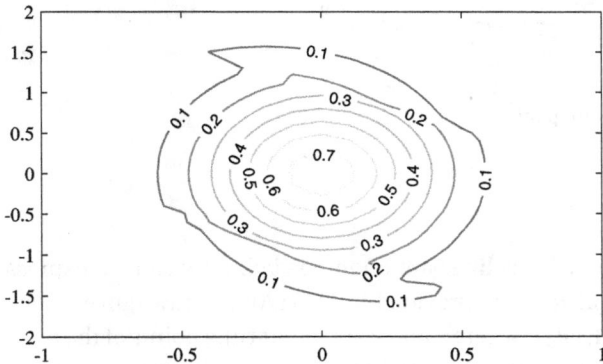

**Figure 7.24:** Contours with height information.

The filled and three-dimensional contours can be drawn with the following statements, as shown in Figures 7.25 and 7.26. In the three-dimensional contour statements, 30 indicates the number of contours, since the default contours are sparsely distributed.

```
>> contourf(x,y,z); figure; contour3(x,y,z,30)
```

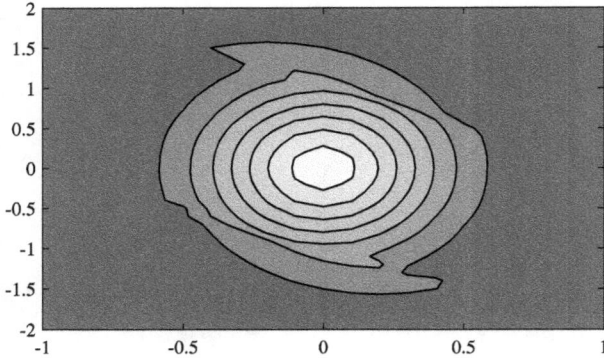

**Figure 7.25:** Filled contours.

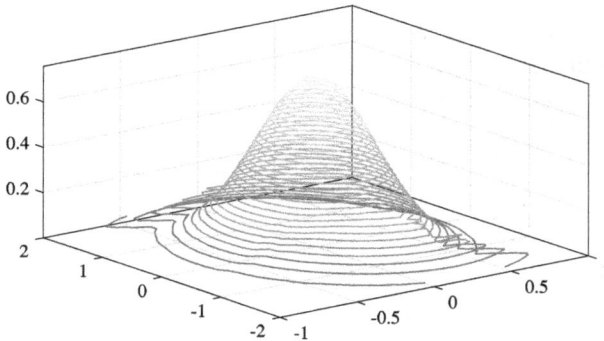

**Figure 7.26:** Three-dimensional contour plots.

## 7.4.2 Quiver plots

In physics courses, the magnetic force lines are often needed. How can we express them in MATLAB? Function `quiver()` is provided in MATLAB to draw quiver plots, with the syntax `quiver(x,y,u_x,u_y)`, where $x$ and $y$ are the starting points of the quivers, while $u_x$ and $u_y$ are the components of the quivers in the $x$ and $y$ axes.

**Example 7.21.** Express the gradient of the function $z = (x^2 - 2x)e^{-x^2-y^2-xy}$ with quiver plots.

**Solutions.** With symbolic computation in MATLAB, the gradient of the function can be computed[22], and the mathematical forms are

$$\frac{\partial z(x,y)}{\partial x} = -e^{-x^2-y^2-xy}(-2x + 2 + 2x^3 + x^2y - 4x^2 - 2xy),$$
$$\frac{\partial z(x,y)}{\partial y} = -x(x - 2)(2y + x)e^{-x^2-y^2-xy}.$$

The gradient of the function is defined as the vector of derivatives of the function with respect to $x$ and $y$. The following statements can be used to draw the gradient as shown in Figure 7.27:

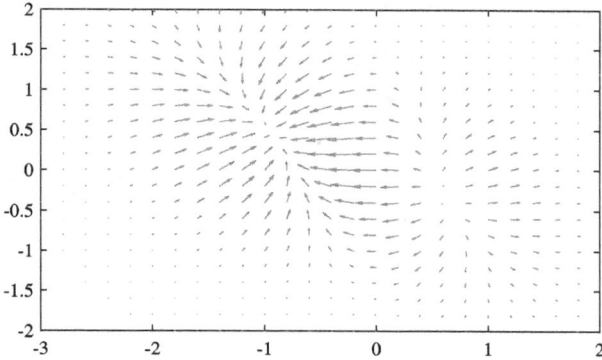

**Figure 7.27:** Gradient of the given function.

```
>> [x,y]=meshgrid(-3:0.2:2,-2:0.2:2); e1=exp(-x.^2-y.^2-x.*y);
   dx=-e1.*(-2*x+2+2*x.^3+x.^2.*y-4*x.^2-2*x.*y);
   dy=-x.*(x-2).*(2*y+x).*e1; quiver(x,y,dx,dy)
```

### 7.4.3 Three-dimensional implicit plots

Functions like `fplot3()` discussed earlier can only be used to draw the curves of three-dimensional explicit functions. If a certain surface is governed by an implicit function $g(x,y,z) = 0$, function `fimplicit3()` in new versions of MATLAB can be used instead, with the syntax

`fimplicit3(fun, [`$x_m$`,`$x_M$`,`$y_m$`,`$y_M$`,`$z_m$`,`$z_M$`]),`

where `fun` can be an anonymous function or a symbolic expression. The ranges of the variables are specified in $x_m$, $x_M$, $y_m$, $y_M$, $z_m$, and $z_M$, with the default values being $\pm 5$. If a set range, $x_m$ and $x_M$, is given, the three axes are assigned the same.

**Example 7.22.** For a surface expressed by the implicit function

$$g(x,y,z) = x\sin(y + z^2) + y^2\cos(x + z) + zx\cos(z + y^2) = 0$$

and the region of interest being $x, y, z \in (-1, 1)$, draw its surface.

**Solutions.** Symbolic expression and anonymous function can both be used to express the original implicit function. With the following statements, the three-dimensional surface of the implicit function can be obtained as shown in Figure 7.28.

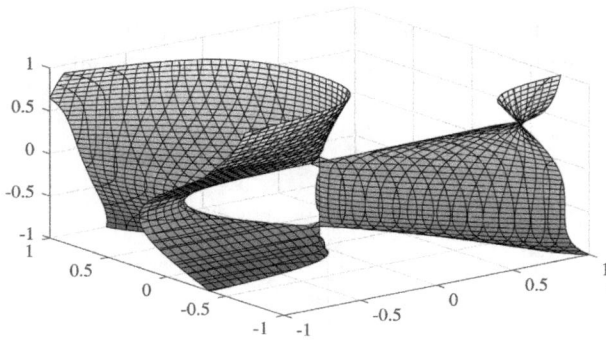

**Figure 7.28:** Surface of a three-dimensional implicit function.

```
>> syms x y z; f=x*sin(y+z^2)+y^2*cos(x+z)+z*x*cos(z+y^2);
   fimplicit3(f,[-2 2]) % 3D implicit function
```

In fact, the implicit function can alternatively be expressed with an anonymous function, and the same results can be obtained. Note that the dot operations should be used whenever possible.

```
>> f=@(x,y,z)x.*sin(y+z.^2)+y.^2.*cos(x+z)+z.*x.*cos(z+y.^2);
   fimplicit3(f,[-1,1])
```

If the following statements are used, a unit sphere $x^2+y^2+z^2 = 1$ can be superimposed, as shown in Figure 7.29.

```
>> f1=x^2+y^2+z^2-1; hold on;
   fimplicit3(f1,[-1 1]); % superimpose a sphere
```

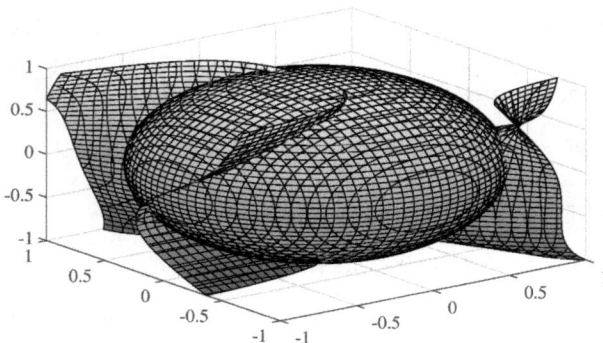

**Figure 7.29:** 3D surface with superimposed sphere.

### 7.4.4 Surfaces of parametric equations

Assume that a three-dimensional function is described by the following parametric equation:

$$x = f_x(u, v), \quad y = f_y(u, v), \quad z = f_z(u, v). \tag{7.4.1}$$

If $u_m \leqslant u \leqslant u_M$, $v_m \leqslant v \leqslant v_M$, function fsurf($f_x$, $f_y$, $f_z$, [$u_m$, $u_M$, $v_m$, $v_M$]) can be called directly to draw the three-dimensional surface, with the default ranges of $u$ and $v$ given as $(-5, 5)$. With earlier versions of MATLAB, function ezsurf() can also be tried to draw the surface.

**Example 7.23.** The famous Möbius strip can be expressed as

$$x = \cos u + v \cos u \cos u/2, \ y = \sin u + v \sin u \cos u/2, \ z = v \sin u/2.$$

If $0 \leqslant u \leqslant 2\pi$, $-0.5 \leqslant v \leqslant 0.5$, draw the surface of the Möbius strip.

**Solutions.** The two symbolic variables $u$ and $v$ should be declared first, and the parametric equations can be entered in MATLAB environment. The following statements can be used to draw the Möbius strip, as shown in Figure 7.30.

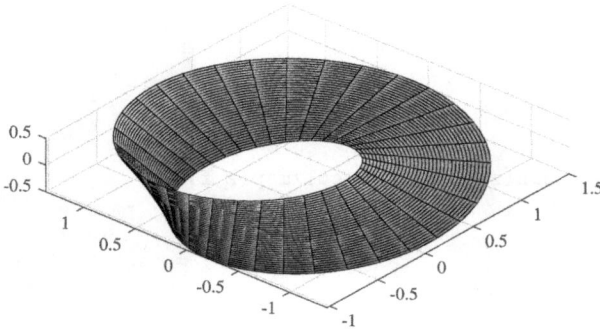

**Figure 7.30:** Möbius strip (after rotation).

```
>> syms u v; x=cos(u)+v*cos(u)*cos(u/2); y=sin(u)+v*sin(u)*cos(u/2);
   z=v*sin(u/2); fsurf(x,y,z,[0,2*pi,-0.5,0.5]) % Möbius strip
```

### 7.4.5 Surfaces of complex functions

Functions surf() and surface() can be used to draw surface plots for given functions. For a complex function $f(z)$ whose independent variable $z$ is complex, it may be rather complicated to draw surfaces in this way, since the original data should be converted into Cartesian coordinates first. Alternatively, the dedicated functions in MATLAB can also be tried, with the syntaxes

$z$=cplxgrid($n$), %generate $(n + 1) \times (2n + 1)$ polar mesh grid matrix
cplxmap($z$,$f$), %compute $f(z)$ and draw the surface of $f$

**Example 7.24.** Draw the surface of the complex function $f(z) = z^4(\sin z + \cos z)$.

**Solutions.** Selecting $n = 40$, the following statements can be used to generate directly the mesh grid matrix. Computing the function, the surface can be drawn as shown in Figure 7.31.

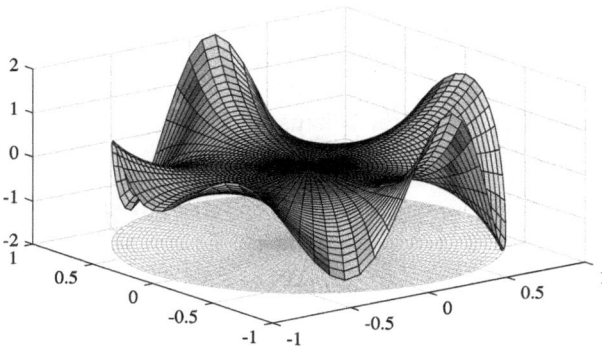

**Figure 7.31:** Surface of complex function.

```
>> z=cplxgrid(40); f=z.^4.*(sin(z)+cos(z)); cplxmap(z,f)
```

### 7.4.6 Spheres and cylinders

The data of the unit sphere centered at the origin can be generated directly with function $[x,y,z]$=sphere($n$). With the function surf(), the surface of the sphere can be drawn immediately. If no argument is returned in the function call, the sphere can be drawn automatically. The argument $n$ in the function call indicates that the sphere is approximated by $n \times n$ faces, and the data is stored in the $(n + 1) \times (n + 1)$ matrices.

**Example 7.25.** Draw a unit sphere centered at the origin first, and then superimpose a smaller one centered at $(0.9, -0.8, 0.6)$, with a radius of $0.3$.

**Solutions.** The data for the unit sphere can be generated first. The data for the smaller sphere can be obtained from the result, so that with the following MATLAB commands, the two spheres are shown in Figure 7.32.

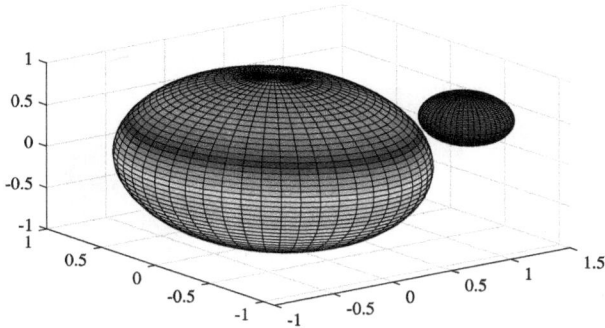

**Figure 7.32:** Sphere surface.

```
>> [x,y,z]=sphere(50); surf(x,y,z), hold on   % unit sphere
   x1=0.3*x+0.9; y1=0.3*y-0.8; z1=0.3*z+0.6; surf(x1,y1,z1)
```

If a curve rotates 360° about the $z$ axis, a cylinder can be traced. If the curve is defined by vector $\boldsymbol{r}$, indicating the radius of the cylinder, the data of the cylinder can be generated with $[\boldsymbol{x},\boldsymbol{y},\boldsymbol{z}]$=cylinder$(\boldsymbol{r},n)$, with the default value of $n$ being 20. If no argument is returned in the function call, the cylinder is drawn automatically. Note that the default interval of $z$ is $z \in (0,1)$.

**Example 7.26.** Assuming that the equation for the cylinder is $r(z) = e^{-z^2/2}\sin z,\ z \in (-1,3)$, draw the cylinder.

**Solutions.** The radius vector can be generated as shown in Figure 7.33.

```
>> z0=-1:0.1:3; r=exp(-z0.^2/2).*sin(z0); plot(z0,r)
```

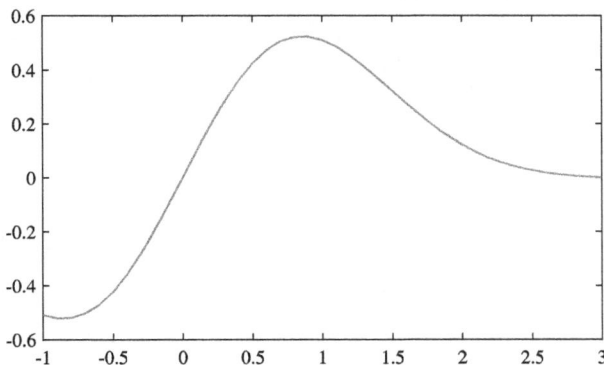

**Figure 7.33:** The radius curve for the cylinder.

If the curve in Figure 7.33 is rotated for 360°, the data for the standard cylinder can be generated. The interval $z \in (0, 1)$ can be mapped into $z \in (-1, 3)$, and the expected cylinder can be drawn as shown in Figure 7.34.

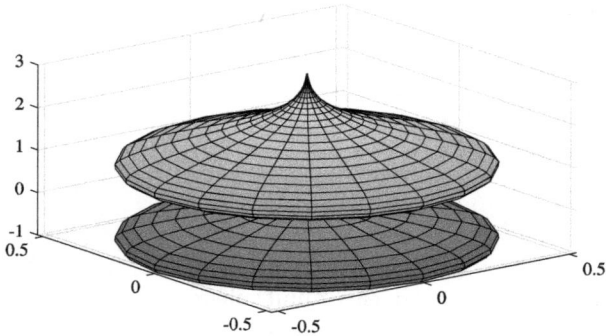

**Figure 7.34:** Surface of the cylinder.

```
>> [x,y,z]=cylinder(r);    % generate the data
   z=-1+4*z;  surf(x,y,z)  % map z from (0,1) to (−1,3)
```

**Example 7.27.** For the given function $y = f(x) = 1 + x \sin 4/x$, and $0 \leqslant x \leqslant \pi$, draw the surface of the curve rotated 360° around the $x$ axis.

**Solutions.** The curve of function $y = f(x)$ can be drawn first as shown in Figure 7.35.

```
>> syms x; f(x)=1+x*sin(4/x); fplot(f,[0,pi])
```

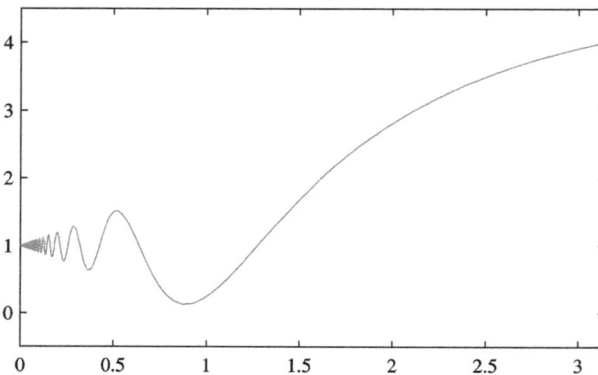

**Figure 7.35:** A two-dimensional curve.

If the curve is rotated, it seems that the function `cylinder()` in MATLAB can be used to draw the surface of the cylinder. The default rotation is to rotate the curve $z$ in the interval $(0, 1)$ around the $z$ axis. In this problem, the $z$ axis is assigned for the values of the function, and rotation should be made about the $x$ axis, then axes should be swapped. Using function `cylinder()`, the final surface can be obtained as shown in Figure 7.36.

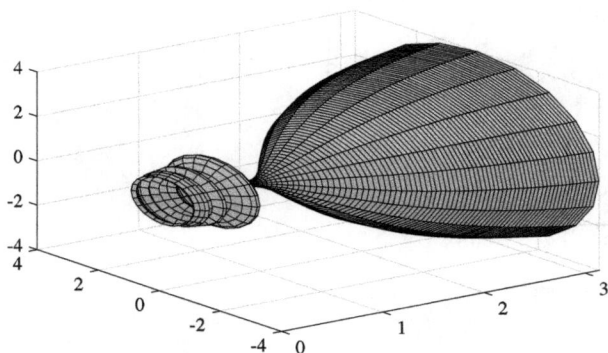

**Figure 7.36:** The three-dimensional surface traced by the curve.

```
>> x=0:0.03:pi; z=1+x.*sin(4./x);        % generate the data
   [x0,y0,z0]=cylinder(z); z0=pi*z0;  % construct the cylinder data
   x1=z0; z1=x0; surf(x1,y0,z1)        % swap axes data and draw
```

### 7.4.7 Voronoi diagrams and Delaunay triangulation

Voronoi diagram and Delaunay triangulation are important concepts in computational geometry. In this section, the fundamental concepts of these diagrams are presented first, and MATLAB solutions to the problems are then discussed.

**Definition 7.2.** Assume that in a two-dimensional plane, there is a set of samples $p_1$, $p_2, \ldots, p_n$, each associated with the variable $R_k$, denoting the range whose distance to $p_k$ is not larger than the distances to any other points. This diagram is referred to as the Voronoi diagram.

If the coordinates of the samples are expressed in vectors $x$ and $y$, MATLAB function `voronoi()` can be used to draw directly the Voronoi diagram, with the syntax `voronoi(x,y)`.

**Example 7.28.** Generate randomly 30 points in the region of the interval $[-1, 1]$. Draw the Voronoi diagram from these points.

**Solutions.** Thirty random points can be generated directly with the following commands, and then the Voronoi diagram from the samples can be drawn directly, as shown in Figure 7.37.

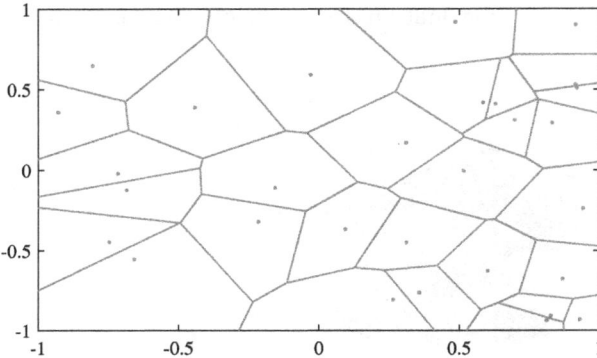

**Figure 7.37:** Voronoi diagram from the samples.

```
>> n=30; x=-1+2*rand(n,1); y=-1+2*rand(n,1);
   voronoi(x,y)
```

**Definition 7.3.** Assuming that there is a set of samples in a plane, stored in vectors *x* and *y*, triangles can be constructed from the points, with no overlapped areas. The triangles are referred to as a triangulation. If the smallest angle in all the triangles is maximized, the triangulation is known as a Delaunay triangulation.

Delaunay triangulation function delaunay() is provided in MATLAB, to compute from the coordinates of *x* and *y* vectors, and the results can be displayed with triplot() function, for Delaunay triangulation. The following statements can be used.

*T*=delaunay(*x*,*y*), triplot(*T*,*x*,*y*)

where the returned argument *T* is a matrix with three columns, with each column being the vertex of a triangle. The function can also be used to handle Delaunay triangulation from three-dimensional samples *x*, *y*, and *z* with *T*=delaunay(*x*,*y*,*z*).

**Example 7.29.** Generate randomly 30 points in a similar way as in Example 7.28, and draw the Delaunay triangulation.

**Solutions.** Thirty random points can be generated, and then with the following statements, the Delaunay triangulation can be drawn, as shown in Figure 7.38.

```
>> n=30; x=-1+2*rand(n,1); y=-1+2*rand(n,1);
   hold on, T=delaunay(x,y); triplot(T,x,y), plot(x,y,'o')
```

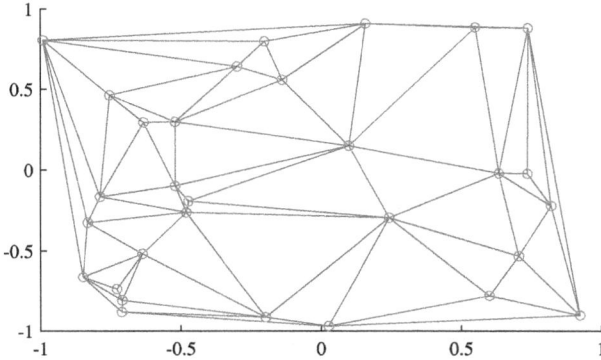

**Figure 7.38:** Delaunay triangulation from random samples.

The function delaunay() provided in MATLAB is of high efficiency, since if 100 000 points are involved, only 0.44 s are needed to complete the Delaunay triangulation, with a total number of triangles being 199 967.

```
>> n=100000; x=-1+2*rand(n,1); y=-1+2*rand(n,1);
   tic, T=delaunay(x,y); toc, size(T)
```

## 7.5 Special treatment of three-dimensional plots

Different surface manipulation facilities are provided in MATLAB. In this section, some special manipulation tactics are presented such as the rotation of surfaces, the conversion methods between different coordinate systems, and surface cut-off and image patching strategies.

### 7.5.1 Rotation of surfaces

As discussed earlier, viewpoints can be set easily. However, the surface themselves are not changed. Only the viewpoints are redefined for observations. Surface rotation facilities are supported in MATLAB, with the function rotate(), having syntax rotate($h$,$v$,$\alpha$), where $h$ is the handle of the surface, which can be returned from the functions such as surf(). Also the handles can be selected in the edit mode of the plots, and then with h=gco command (get current object) to extract the handle. Vector $v$ is used as the baseline for the rotation, which is a $1 \times 3$ vector, storing a point in a three-dimensional space. The baseline is defined as the segment between the point and the origin. $\alpha$ is the angle of rotation in degrees. To rotate about the positive direction of the $x$ axis, one has to set $v$=[1,0,0]. If the negative direction of $y$ axis is used, then $v$=[0,-1,0].

**Example 7.30.** Consider again the piecewise function studied in Example 4.20. Rotate the surface of the function.

**Solutions.** The following statements can be used to draw the three-dimensional surface of the piecewise function, shown in Figure 7.19.

```
>> [x,y]=meshgrid(-1:.04:1,-2:.04:2);  % mesh grid matrix
   z=0.5457*exp(-0.75*y.^2-3.75*x.^2-1.5*x).*(x+y>1)+...
     0.7575*exp(-y.^2-6*x.^2).*((x+y>-1) & (x+y<=1))+... % piecewise
     0.5457*exp(-0.75*y.^2-3.75*x.^2+1.5*x).*(x+y<=-1);
   h=surf(x,y,z);  % draw surface and return the handle
```

Apart from drawing the surface, the handle h of the surface can be obtained. If the baseline for rotation is the original $x$ axis, and the rotation is by 15° in the counterclockwise direction, the following statements should be used, to get the picture as shown in Figure 7.39.

```
>> rot_ax=[1,0,0]; rotate(h,rot_ax,15)  % rotate 15° about x
```

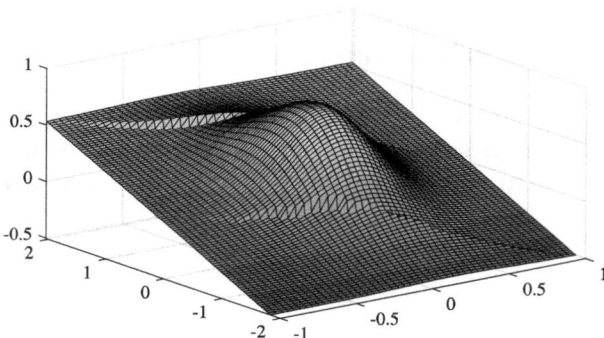

**Figure 7.39:** Rotation by 15° about the $x$ axis.

If we want to rotate the surface about the line between the origin and the spatial point $(1, 1, 1)$ by 15°, the following statements can be used, and the result is shown in Figure 7.40. Note that here h should be drawn and defined again, and the previous one cannot be used, otherwise the rotation will be applied to the surface shown in Figure 7.39.

```
>> h=surf(x,y,z); rot_ax=[1,1,1]; rotate(h,rot_ax,15)  % rotate 15°
```

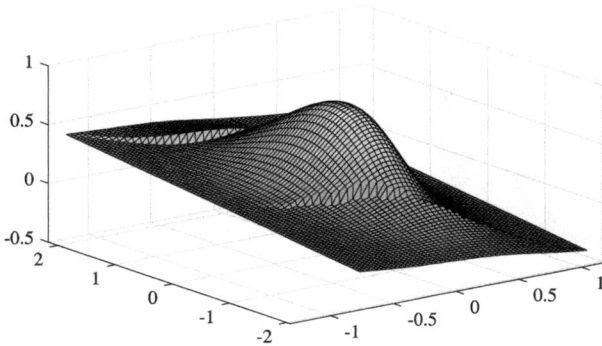

**Figure 7.40:** Rotate by 15° about the line between the spatial point $(1, 1, 1)$ and the origin.

A loop structure can be used to rotate the original surface relative to the $x$ axis in animation form, rotating by 1° with a 0.02 s pause. Here the command `axis tight` is used to keep the scales of the axes unchanged in the rotation process. It is worth mentioning that the angle in the rotation is 1 rather than $i$, since a rotation by 1° in the loop is applied to the previous result.

```
>> h=surf(x,y,z); r_ax=[1 0 0]; axis tight
   for i=0:360, rotate(h,r_ax,1); pause(0.02), end
```

### 7.5.2 Axis specification for surfaces

The surface plots of the function $f(x, y)$ were discussed so far. If a function $g(x, z)$ is given, how can we draw a three-dimensional surface plot? There is no direct function in MATLAB to draw such plots. The rotation method presented earlier can be used to draw the surface, but the angle for rotation cannot be computed in an easy manner. An alternative method can be used to draw the surface of the function.

**Example 7.31.** Draw the surface of the function $y(x, z) = (x^2 - 2x)e^{-x^2-z^2-xz}$.

**Solutions.** Letting $\hat{x} = x$, $\hat{z} = y$, $\hat{y} = z$, the original function $y(x, z)$ can be rewritten as

$$\hat{z}(\hat{x}, \hat{y}) = (\hat{x}^2 - 2\hat{x})e^{-\hat{x}^2-\hat{y}^2-\hat{x}\hat{y}},$$

and the mesh grid matrices can be constructed as $\hat{x}$, $\hat{y}$, $\hat{z}$. Then the function $\text{surf}(\hat{x}, \hat{y}, \hat{z})$ can be used to draw a three-dimensional surface, with $\text{surf}(x, z, y)$. Employing the following statements, the expected surface can be drawn, as shown in Figure 7.41.

```
>> [x,z]=meshgrid(-3:0.1:2,-2:0.1:2);   % mesh grid matrices
   y=(x.^2-2*x).*exp(-x.^2-z.^2-x.*z); surf(x,z,y)
```

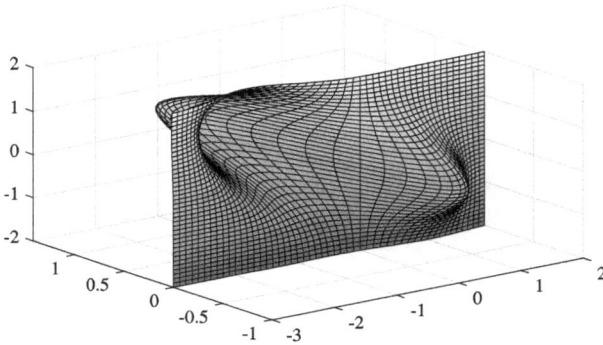

**Figure 7.41:** Surface of $f(x, z)$ function.

It can be seen from the example that if the function is not $f(x, y)$, one can swap the axes so that the correct surface can be found.

### 7.5.3 Cutting of surfaces

In drawing MATLAB plots, if the function value at a certain point is not numeric, which can be NaN or Inf, the point is not displayed. With this characteristics, the region in a plot or a surface can be cut off.

**Example 7.32.** Consider the function with two independent variables, $z = f(x, y) = \sin(xy^2/20)$. Letting $x^2 + y^2 \leqslant 9$, $2x + 3y \leqslant 2$, draw the surface.

**Solutions.** The question now is how to draw a portion of a surface? The mesh grid in the interval $(-3, 3)$ can be generated first, and then we can compute the matrix $z$. For the two inequalities, if the values of the mesh grid points which do not satisfy the inequalities are set to NaN, the surface in the expected region can be obtained, as shown in Figure 7.42.

```
>> [x,y]=meshgrid(-3:0.1:3); z=sin(x.*y.^2/20);
   z(x.^2+y.^2>9)=NaN; z(2*x+3*y>2)=NaN; surf(x,y,z)
```

### 7.5.4 Patches in surfaces

Images can be patched onto the surface, the color image should be converted first into an indexed one, with the syntax $[X, \text{map}]=\text{rgb2ind}(W, n)$, where $W$ is a three-dimensional array containing the original RGB image. After conversion, $X$ is a homochromous bitmap matrix, while map is an indexed table for the color, with $n$ rows and three columns. Color-indexed table image is one of the commonly used color images. The color of each pixel can be restored by the color index table. If the color im-

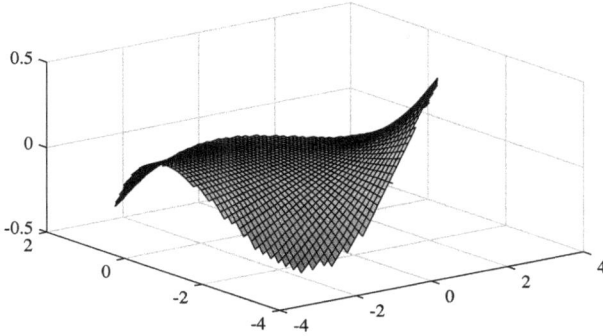

**Figure 7.42:** Three-dimensional surface in the expected region.

age is represented by a matrix and indexed table, the storage and color resolution is smaller than that of the RGB image.

Matrix $X$ can be converted into a double precision matrix, with function `surface()` showing a three-dimensional surface of the function, with the syntaxes given by

```
surface(x,y,z,X,'FaceColor','texturemap','EdgeColor','none')
colormap(map), grid % color index map restoration
```

where $x$, $y$, $z$ are the mesh grid matrices, while matrix $X$ is the image file information. The patching method of the surfaces are made through examples.

**Example 7.33.** Patch the "tiantan" image to the surface in Example 7.9.

**Solutions.** The "tiantan" image can be read into MATLAB workspace first, and then converted into a color-indexed image. Drawing the three-dimensional surface, the indexed image can be patched to the surface, and finally, the color can be restored. The processed result can be obtained as shown in Figure 7.43.

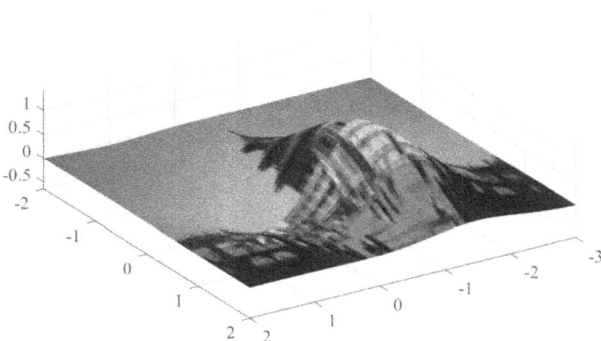

**Figure 7.43:** A patched three-dimensional surface.

```
>> W=imread('tiantan.jpg'); [X,map]=rgb2ind(W,128);
   X=double(X); [x,y]=meshgrid(-3:0.1:2,-2:0.1:2);
   z=(x.^2-2*x).*exp(-x.^2-y.^2-x.*y);
   surface(x,y,z,X,'FaceColor','texturemap','EdgeColor','none')
   colormap(map), grid
```

## 7.6 Four-dimensional plots

In the previous sections, the presentations were made upon the assumption that the function is defined in the form of $z = S(x, y)$ in the three-dimensional space. If a mathematical function with three independent variables is given by $v = V(x, y, z)$, a volume visualization plot can be obtained. There are several applications, such as the temperature inside a solid body, flow in the fluid stream, concentration of certain solutions, and so on. Such plots cannot be drawn with regular three-dimensional graphics facilities. It is not possible to draw four-dimensional plots directly either. Special three-dimensional spatial plots can be obtained, and slices can be made to observe the internal information of the three-dimensional objects. This method is also known as the volume visualization technique. Computer tomography (CT) is a good example to show the internal structure of a solid object with slices.

In this section, slicing method is illustrated first, then a graphical user interface by the author is presented in implement volume visualization easily. Finally, the making and playing of animation videos are illustrated.

### 7.6.1 Slices

Three-dimensional mesh grid data can be generated with function `meshgrid()`, and with three-dimensional arrays $x$, $y$, and $z$, the volume data $V$ can be computed with the dot operation. Function `slice()` can then be used to draw slices, with the syntax `slice(x,y,z,V,x_1,y_1,z_1)`, where $x$, $y$, $z$, and $V$ are volume data, and variables $x_1$, $y_1$, $z_1$ can be used to describe the slices. If the slices are described by constant vectors, the slices perpendicular to the axes can be made. The slices can also be assigned as slope planes or surfaces. Examples are illustrated to demonstrate the methods used.

**Example 7.34.** A function with three independent variables is given below as

$$V(x, y, z) = \sqrt{x^x + y^{(x+y)/2} + z^{(x+y+z)/3}}.$$

Observe the function with the volume visualization technique. Slices can be used to observe the properties of the function.

**Solutions.** Since the square roots are involved, the variables $x$, $y$, and $z$ are defined as nonnegative values. Mesh grid data can be generated with the following statements, the volume data **V** can be computed with the dot operation. The slices perpendicular to the axes can be assigned. For instance, a set of slices at $x = 1$ and $x = 2$ can be defined. Also, the slices at $y = 1$, $y = 2$, $z = 0$, and $z = 1$ can be defined as well. The slice plots can be obtained as shown in Figure 7.44.

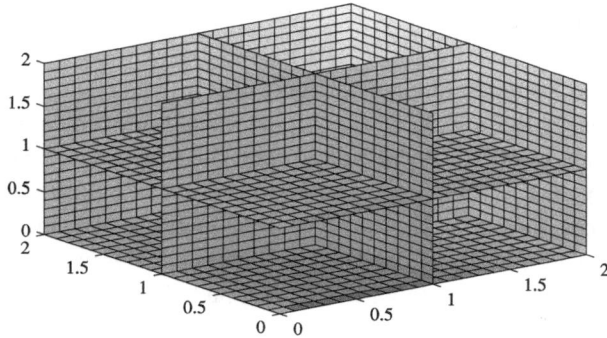

**Figure 7.44:** Slice with respect to the horizontal plane.

```
>> [x,y,z]=meshgrid(0:0.1:2);           % 3D mesh grid data
   V=sqrt(x.^x+y.^((x+y)/2)+z.^((x+y+z)/3)); % compute function
   slice(x,y,z,V,[1 2],[1 2],[0 1]); % draw plots with slices
```

Similar to the method discussed earlier, the plane $z = 1$ can be generated first, then a 45° rotation along the $x$ axis can be made. The slice data can be extracted as $x_1$, $y_1$, and $z_1$. The volume visualization plot with slices can be drawn with function $\text{slice}()$, as shown in Figure 7.45.

```
>> [x0,y0]=meshgrid(0:0.1:2); z0=ones(size(x0)); % the z = 1 plane
   h=surf(x0,y0,z0); rotate(h,[1,0,0],45);       % rotate 45° along x
```

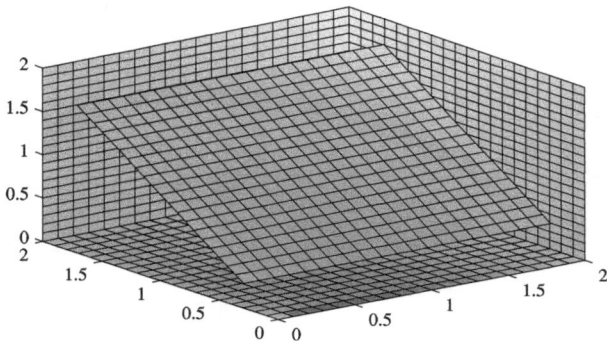

**Figure 7.45:** Arbitrary slope plane.

```
x1=get(h,'XData'); y1=get(h,'YData');
z1=get(h,'ZData'); slice(x,y,z,V,x1,y1,z1)    % extract data
hold on, slice(x,y,z,V,2,2,0)                  % draw slices
```

### 7.6.2 A volume visualization interface

In order to visualize the slices in a simpler manner, a simple user interface vol_visual4d() is written. Before using the interface, the volume visualization data **x**, **y**, **z**, and **V** should be presented in MATLAB workspace, then the function can be called with vol_visual4d(**x**,**y**,**z**,**V**). The controls in the interface can be used to handle the positions of the slices. In this section, the use of the interface is demonstrated, and the programming of graphical interface will be illustrated later in Chapter 10. It is recommended to read the program when you complete learning Chapter 10.

**Example 7.35.** With the data generated in Example 7.34, function vol_visual4d() can be used to initiate the interface, and the slice display of the data is shown in Figure 7.46. The scroll bars provided in the interface can be used to adjust the positions of the slices. The checkboxes on/off can be used to turn on and off the slices in cer-

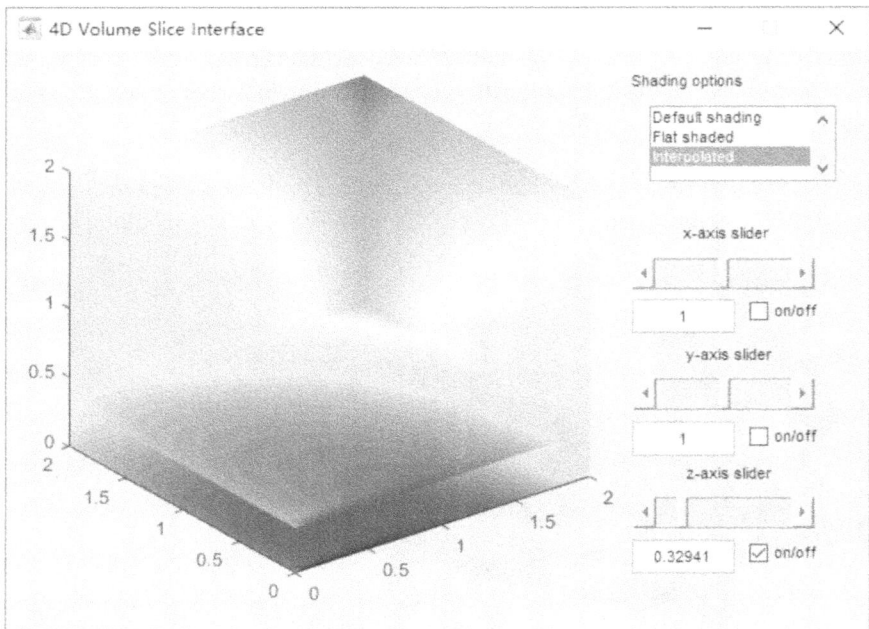

**Figure 7.46:** The slice view interface.

tain axes. The listbox Shading options allows the user to select a shading mode in the volume visualization results.

```
>> [x,y,z]=meshgrid(0:0.1:2);
   V=sqrt(x.^x+y.^((x+y)/2)+z.^((x+y+z)/3)); % generate data
   vol_visual4d(x,y,z,V);                     % view the slices
```

### 7.6.3 Creating and playing of three-dimensional animations

Three-dimensional animation can also be understood as a four-dimensional graphics, where the fourth dimension is time. For a given three-dimensional surface, if it changes with time, each surface can be extracted from getframe() function to get the handle. A series of handles can be extracted in this way, and then function movie() can be used to generate the animation. Examples are given in this section to show the video making and playing processes.

**Example 7.36.** Consider a time-varying function $z(x, y, t) = \sin(x^2 t + y^2)$. Make an animation for the three-dimensional surface varying with time $t$.

**Solutions.** The processing of a three-dimensional animation can be divided into two steps, the first being the movie-making process. At any time instance, the three-dimensional surface of the function can be drawn. Then, with getframe() function, the handle of the image frame can be generated. A series of handles can be obtained with this method. In order to make the animation look smooth, the function axis() can be used to fix the axis in a certain range. The second step is the play of animation. With the series of handles, function movie() can be used to play movies. With the following statements, the expected animation results can be completed:

```
>> t=linspace(0,2*pi); [x,y]=meshgrid(-2:0.1:2);
   for i=1:length(t)  % process each instance in loop structure
       z=sin(x.^2*t(i)+y.^2); surf(x,y,z);     % draw 3D plot
       axis([-2,2,-2,2,-1,1]); h(i)=getframe; % extract a frame
   end
   figure, movie(h)   % direct play of the movie
```

It should be pointed out that function getframe() can be used to extract a three-dimensional frame. For two-dimensional plots, the function can also be used to extract a frame. Therefore, the function can also be used to make movies for two-dimensional plots. Besides, VideoWriter() function can be used to open a movie file, while function writeVideo() can be used to write a frame into a movie file. The movie animation is demonstrated in the following example.

**Example 7.37.** Two-dimensional Brownian motion in Example 6.19 is animated and can be made into a movie.

**Solutions.** Assume that 200 steps in Brownian motion are to be demonstrated. The following statements can be used to generate data for the animation. The video can be made with the data, and a video file can be created as brown.avi, which can be played by any multimedia player.

```
>> n=30; x=randn(1,n); y=randn(1,n); sig=0.3;    % initial position
   figure(gcf), hold off;  % set the current window
   h=plot(x,y,'o'); axis([-30,30,-30,30])
   vid=VideoWriter('brown.avi'); open(vid); % open movie file
   for k=1:200, % animation of moving 200 steps
       x=x+sig*randn(1,n); y=y+sig*randn(1,n);
       set(h,'XData',x,'YData',y), drawnow    % make a frame
       hVid=getframe; writeVideo(vid,hVid);   % write a frame
   end
   close(vid)    % close movie file
```

## 7.7 Problems

7.1   If the position of a spatial particle can be described as $x(t) = \cos t + t \sin t$, $y(t) = \sin t - t \cos t$, $z(t) = t^2$, and $t \in (0, 2\pi)$, draw the moving trajectory of the particle, and also show dynamically the animated trajectory of the particle.

7.2   Draw the surfaces of the functions $xy$, $\sin xy$, and $e^{2x/(x^2+y^2)}$.

7.3   Draw the surfaces of the function $f(x,y) = \sin\sqrt{x^2+y^2}/\sqrt{x^2+y^2}$, with $-8 \leqslant x, y \leqslant 8$.

7.4   A vertical cylinder can be described by the parametric equations $x = r \sin u$, $y = r \cos u$, $z = v$. The radius is $r$. If the axes $x$ and $z$ are swapped, the cylinder about the $x$ axis can be obtained. Draw in the same coordinates the cylinders using different axes and radii.

7.5   Draw the three-dimensional surface for the following parametric equations[11]:
(1) $x = 2\sin^2 u \cos^2 v$, $y = 2\sin u \sin^2 v$, $z = 2\cos u \sin^2 v$, $-\pi/2 \leqslant u, v \leqslant \pi/2$;
(2) $x = u - u^3/3 + uv^2$, $y = v - v^3/3 + vu^2$, $z = u^2 - v^2$, $-2 \leqslant u, v \leqslant 2$.

7.6   Assume that the vertex of a cone is at $(0, 0, 2)$, and its base is at $z = 0$, with a the radius of the circle being 1. Draw the surface of the cone.

7.7   Draw the surfaces and contour lines of the following functions. Functions waterfall(), surfc(), and surfl() can be used to observe the results.
(1) $z = xy$,   (2) $z = \sin x^2 y^3$,
(3) $z = (x-1)^2 y^2/[(x-1)^2 + y^2]$,   (4) $z = -xy\,e^{-2(x^2+y^2)}$.

7.8   Draw the surface of the implicit function

$$(x^2 + xy + xz)e^{-z} + z^2 yx + \sin(x + y + z^2) = 0.$$

7.9   Draw the two surfaces of the given two functions $x^2 + y^2 + z^2 = 64$ and $y + z = 0$
       and observe their intersections.

7.10  With the known functions

$$x = u \sin t, \quad y = u \cos t, \quad z = t/3, \quad t \in (0, 15), \ u \in (-1, 1)$$

       draw the surface.

7.11  In MATLAB graphics, if the value of the function at a certain point is NaN, the
       point is not displayed. Draw the surface of $z = \sin xy$, and cut off the regions
       when $x^2 + y^2 \leqslant 0.5^2$.

7.12  Draw the surface of $x(z, y) = (z^2 - 2z)e^{-z^2 - y^2 - zy}$.

7.13  If $x = \cos t(3 + \cos u)$, $\sin t(3 + \cos u)$, $z = \sin u$, with $t \in (0, 2\pi)$, $u \in (0, 2\pi)$, draw
       the surface of the function.

7.14  Draw the surface of the complex function $f(z) = e^{-z} \sin(z^2)$.

7.15  Function cplxroot() is provided in MATLAB to draw the surface of $\sqrt[n]{z}$. Draw
       the surfaces of $\sqrt[3]{z}$ and $\sqrt[4]{z}$.

7.16  For the following functions with three independent variables, draw the slice
       plots with the volume visualization technique:

       (1) $V(x, y, z) = \sqrt{e^x + e^{(x+y)-xy} + e^{(x+y+z)/3 - xyz}}$,   (2) $V(x, y, z) = e^{-x^2 - y^2 - z^2}$.

7.17  For the given complex function $f(z, t) = z^4(\sin zt + \cos zt)$, make an animation
       for $t \in (0, 2\pi)$ for the surface of the function, and create a standard video file
       cplxsur.avi, which can be played with Windows Media Player or any other
       video player.

# 8 MATLAB and its interface to other languages

MATLAB programming is simple and of high efficiency. In some certain cases, we are also interested in solving problems with low-level functions. For instance, since MATLAB is a descriptive language, the code with large loop structures is very slow. The slow piece of code can be implemented instead with languages such as C. The concise and practical characteristics of MATLAB can be adopted so as to improve the high effectiveness of the programs.

In the following cases, the interface of MATLAB with C language should be used:

(1) As illustrated earlier, in some special computation, since the speed of the explanatory computer languages such as MATLAB is usually very low, the programs written in C language may speed up the whole problem solving to avoid bottleneck phenomena. C language can be considered to rewrite certain parts of the program, with the C Mex technique.

(2) Before using MATLAB language to write programs, there may be a lot of existing code written in C, Fortran, or other languages. The user wants to use the code directly rather than rewrite it all. An interface should be written to load MATLAB variables into C code, then the C code can be executed and the output should be written back to MATLAB environment. Repetitive and laborious work may be avoided by using this code.

(3) When low-level device should be accessed, such as to exchange information with hardware interfaces and devices, MATLAB cannot be used alone. Low-level code in C language should be used to implement the facilities. Of course, if equipped with adequate toolboxes and hardware devices, MATLAB can be used to implement the facilities in a simple and standardized manner.

(4) For security reasons, in one wants to hide the source code, and the pseudocode format is not used, the kernel part of the code can be written in C. Executable files can be supplied to MATLAB users. Protection of the source code can then be improved.

Apart from the calls to C programs from MATLAB, its functions can also be called within C programs. The programming styles are made more flexible. For instance, dynamic data exchange (DDE) in Microsoft Windows can be used, and also the ActiveX techniques can be easily adopted in the files.

The topics in this chapter are arranged as follows. In Section 8.1, the C variable and function support for MATLAB are introduced. The information exchange methods for various data styles between MATLAB and C are demonstrated. In Section 8.2, manipulations of various input and returned arguments are presented. Reading and writing of MAT files are also discussed. The calling of MATLAB functions from within C programs are presented in Section 8.3. In Section 8.4, the direct translation of MATLAB functions

https://doi.org/10.1515/9783110666953-008

to C based standalone executable files are addressed, such that the programs can be executed without the support of MATLAB.

## 8.1 Introduction to C interfaces with MATLAB

In MATLAB, an interface to C language written in Mex format can be called. MATLAB functions can also be called within C programs. A series of interfaces to C language from MATLAB are provided. The programs can be called from C language files.

According to the Mex format, C language files may be written through a specific compiler, and the call to the files can be made from MATLAB. The syntaxes are the same as if one was calling MATLAB functions, from the users' viewpoint. In this section, the setting of C compilers is introduced first. Then the data types supported in Mex format are presented. Finally, the structure and programming of Mex programs are demonstrated with examples.

### 8.1.1 Environment setting of compilers

In recent versions of MATLAB, the TDM-GCC MinGW compiler for C is provided, and it is also downloadable from

```
https://sourceforge.net/projects/tdm-gcc/?source=typ_redirect
```

After download, it can be installed in the default path. The following command can be used to automatically set the compiler environment in MATLAB. Note that every time MATLAB is invoked, the command should be called. The user may also write the command into the startup.m file.

```
>> setenv('MW_MINGW64_LOC','C:\TDM-GCC-64')
```

To compile C programs, the following command should be used:

```
>> mex filename.c
```

If the compilation process is successful, an executable file "filename.mexw64" can be generated. The syntax of the executable functions can be called in the same way as other regular MATLAB functions. Later, the function calls to Mex files assume that the Mex programs are already compiled.

### 8.1.2 Data types in Mex

For simplicity, the unified data type is defined in C for MATLAB, namely MATLAB array type. This data type does not lose generality. It can be used to describe scalars, vectors, and matrices, or it can be used for structured data, cells, and so on. In C language, a MATLAB variable **A** is defined in a unified form as `mxArray *A`.

For string variables, it can also be defined as `mxChar`. The data type can be identified directly in C language as:

(1) **Detect the data type of an input argument.** The type of a variable can be found with the following statements:

```
mxClassID k_id=mxGetClassID(mxArray *ptr)
```

For the pointer `ptr` of an `mxArray` object, the returned argument `k_id` is the data type of the variable. It is an `mxClassID` variable in Mex format. The commonly used identifiers in C Mex format are listed in Table 8.1.

**Table 8.1:** The supported identifiers in MATLAB.

| identifier | data type | class |
|---|---|---|
| mxDOUBLE_CLASS | double precision | 'double' |
| mxINT8_CLASS | 8 bit integer | 'int8' |
| mxINT16_CLASS | 16 bit integer | 'int16' |
| mxINT32_CLASS | 32 bit integer | 'int32' |
| mxCHAR_CLASS | string | 'char' |
| mxCELL_CLASS | cell array | 'cell' |
| mxSINGLE_CLASS | single precision | 'single' |
| mxUINT8_CLASS | 8 bit unsigned | 'uint8' |
| mxUINT16_CLASS | 16 bit unsigned | 'uint16' |
| mxUINT32_CLASS | 32 bit unsigned | 'uint32' |
| mxSTRUCT_CLASS | structured variable | 'struct' |
| mxUNKNOWN_CLASS | unknown | – |

(2) **Get the total number of elements in an input argument.** The total number of elements in a variable pointed by `ptr` can be measured by

```
int n=mxGetNumberOfElements(mxArray *ptr),
```

where $n$ is the number of elements in the variable. It is equivalent to the MATLAB command `prod(size(A))` for variable **A**.

(3) **Measure the size of input arguments.** The size of the argument with pointer `ptr` can be measured with the function

```
int m=mxGetNumberOfDimensions(mxArray *ptr);
```

(4) **Test whether a variable is of certain data type.** For instance,

```
bool k=mxIsChar(mxArray *ptr)
```

can be used to test whether the variable pointed by `ptr` is a string or not. If it is, then the function returns 1, otherwise it returns 0. The returned argument is a logic one. Similar functions are `mxIsCell()`, `mxIsClass()`, `mxIsNaN()`, and so on. The meanings of the functions are straightforward, and they are not described here.

The following suffixes are used for the function names for C interface to MATLAB:

- Suffix `mx`. Such functions are used to create or read MATLAB variables. They can be used to manipulate directly MATLAB arrays defined as `mxArray`. They are declared in the header file `matrix.h`. In C programming, it must be included with command `#include "matrix.h"`.
- Suffix `mat`. It can be used for the functions to access the MAT data files. The functions are defined in the header file `mat.h`.
- Suffix `mex`. It is generally used for calling MATLAB functions from C programs. The functions are defined in the header file `mex.h`.
- Suffix `eng`. It is for C functions to call MATLAB computational engines. They are defined in the header file `engine.h`.

When calling a certain function, the header files should be declared with `#include` command, for instance,

```
#include "mex.h"
```

The above data structures and functions are defined in the header files. With such declarations, the MATLAB data types can be accessed with the C functions.

### 8.1.3 Mex file structures

The fundamental Mex file structure is shown in Figure 8.1. With this structure, an entrance function in Mex is defined. The pointers to the input arguments can be retrieved, such that the input arguments can be loaded into a C program. The major body of C language can be used and the result can be written back to the caller of MATLAB. The main entrance of the Mex function is led by the `mexFunction()` statement. The syntax of the function is a fixed one:

```
void mexFunction(int nl,mxArray *pl[],int nr,const mxArray *pr[])
```

where `nl` and `nr` are the numbers of returned and input arguments to the MATLAB environment, which are equivalent to `nargout` and `nargin`. The variables `*pl[]` and `*pr[]` are respectively the pointers of returned and input arguments. For instance, `*pr[0]` is the pointer of the first input argument, `*pr[1]` is the pointer of the second input argument, and so on. These variables are equivalent to the variables `varargin` and `varargout`. In the function call, the pointers of the three variables `nl`, `nr`, and `*pr[]` are automatically assigned. The pointer `*pl[]` should be assigned dynamically. The following C functions are commonly used in Mex programming:

(1) Retrieve the numbers of rows and columns of a matrix. The facilities can be implemented with the functions `mxGetM()` and `mxGetN()`. For instance, the size of the

```
┌─────────────────────────────────────────────┐
│           Mex header, mex.h                   │
└─────────────────────────────────────────────┘
                      ↓
┌─────────────────────────────────────────────┐
│ leading void mexFunction(int nl, mxArray *pl[],│
│        int nr, const mxArray *pr[])           │
└─────────────────────────────────────────────┘
                      ↓
┌─────────────────────────────────────────────┐
│        set input argument with pr[]           │
└─────────────────────────────────────────────┘
                      ↓
┌─────────────────────────────────────────────┐
│      input arguments, call the function       │
│       pl[]=mxCreateDoubleMatrix()             │
└─────────────────────────────────────────────┘
                      ↓
┌─────────────────────────────────────────────┐
│             main body of C                    │
└─────────────────────────────────────────────┘
                      ↓
┌─────────────────────────────────────────────┐
│     write returned arguments to pl[]          │
└─────────────────────────────────────────────┘
                      ↓
┌─────────────────────────────────────────────┐
│                 return                        │
└─────────────────────────────────────────────┘
```

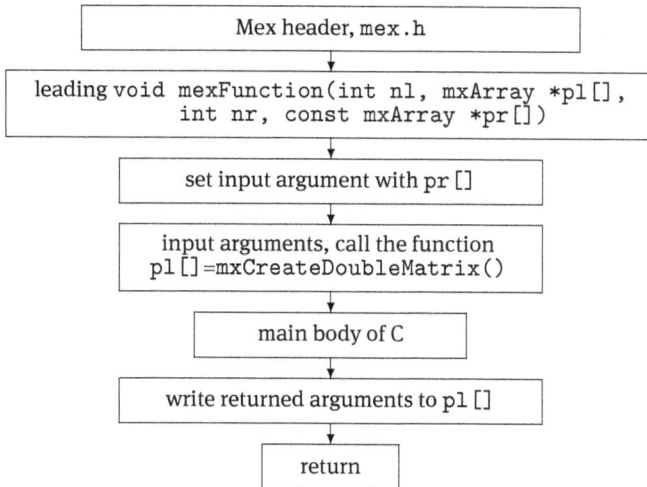

**Figure 8.1:** Illustration of Mex file structures.

$(k + 1)$th input argument can be measured from `mxGetM(pr[k])` and `mxGetN(pr[k])`. This function is, in fact, a simpler form of `mxGetDimensions()` for matrix problems. The reason for the correspondence of $k + 1$ and $k$ is that the subscripts in C language start from 0, rather than 1.

(2) Retrieve the pointer of matrix variables. The pointer of a variable can be retrieved with `mxGetPr()` function. For instance, the pointer of the $(k + 1)$th input argument can be retrieved from `mxGetPr(pr[k])`. If the $(k + 1)$th returned variable is successfully allocated, function `mxGetPr(pl[k])` can be used to retrieve its pointer. If the $(k + 1)$th variable is a scalar, the function calls `mxGetM()` and `mxGetN()` can be omitted, and `mxGetPr()` function can be changed to `mxGetScalar(pr[k])`. It is worth mentioning that, even though the input argument is a matrix, it can still be expressed as a vector in C language, which is the expansion of the elements in the original matrix in a column-wise manner.

(3) **Test whether a matrix is complex or not.** The `mxIsComplex(pr[k])` function can be used to judge whether the $(k + 1)$th input argument has imaginary part or not. If imaginary part is present, the returned result is 1, otherwise is 0. The pointer of the imaginary part of a complex matrix can be retrieved with the function `mxGetPi(pr[k])`.

(4) **Dynamical assignment of output argument pointer.** The following statement can be used to allocate a matrix, and the pointer can be obtained:

```
pl[k]=mxCreateDoubleMatrix(mrows,ncols,mxREAL);
```

where `mrows` and `ncols` are the numbers of rows and columns of the newly created matrix. The constant `mxREAL` indicates a real matrix, while `mxCOMPLEX` indicates a complex matrix. With such a function, the memory space of the $(k + 1)$th returned variable is allocated, whose pointer can be retrieved with function mxGetPr(). In the above function, the pointers of the arguments are represented by pl and pr. In fact, the use of the functions is not limited to fixed pointers. It applies to any mxArray MATLAB data types.

(5) Functions mexGetVariable() and mexPutVariable() can be used in Mex files to exchange information between the function and MATLAB workspace.

**Example 8.1.** Write a Mex function $C$=mex_ex81($A$,$B$) to evaluate the product of two matrices in C language.

**Solutions.** If a Mex C program is expected, the matrices $A$ and $B$ should be loaded first into the C program, then a triple-loop structure can be used to compute their product. Finally, the result in C should be written back to MATLAB environment. The procedures in C code can be considered as follows. The two input arguments $A$ and $B$ can be passed to the C program through pointers pr[0] and pr[1]. The product of the two matrices can be computed with the subfunction mat_multiply, and the result can be returned to variable $C$. Function mxCreateDoubleMatrix() can be used to create a blank matrix, and its pointer can be returned with the function mxGetPr(). The following C code can be written:

```c
#include "mex.h"
void mat_multiply(double *A, double *B, double *C,
        int mA, int nA, int mB, int nB)
{int i,j,k,m=0;
    for (i=0; i<mA; i++){for (j=0; j<nB; j++){C[j*mA+i]=0;
        for (k=0; k<mB; k++){C[j*mA+i]+=A[k*mA+i]*B[j*mB+k];
}}}}
/* main program */
void mexFunction(int nl, mxArray *pl[], int nr, const mxArray *pr[])
{double *Ap, *Bp, *Cp; int mA,nA,mB,nB,mC,nC;
    Ap=mxGetPr(pr[0]); Bp=mxGetPr(pr[1]);
    mA=mxGetM(pr[0]); nA=mxGetN(pr[0]);
    mB=mxGetM(pr[1]); nB=mxGetN(pr[1]);
    pl[0]=mxCreateDoubleMatrix(mA,nB,mxREAL);
    Cp=mxGetPr(pl[0]);
    mat_multiply(Ap, Bp, Cp, mA, nA, mB, nB);
}
```

where the entrance of the function is mexFunction(). In the function, the pointers of matrices $A$ and $B$ can be extracted, and then their product can be computed with function mat_multiply(). Creating a pointer for the output argument, the result can be written automatically to the pointer. All the procedures should be given in the C program.

With the following command, the program can be compiled:

```
>> mex mex_ex81.c
```

so that the executable file `mex_ex81.mexw64` is generated after compilation. The following MATLAB commands can be used to compute the product of the two matrices $A^T$ and $B$:

```
>> A=[1 2 3; 4 5 6]; B=[1 2; 3 4];
   C=mex_ex81(A',B), D=A'*B, E=mex_ex81(A,B), F=A*B
```

The results obtained are as follows. It can be seen that the syntax is similar to ordinary MATLAB functions, where matrices $C$ and $D$ are identical. If one wants to multiply $A$ and $B$ directly, since they are not compatible, $F$ matrix cannot be obtained. However, function `mex_ex81()` can still be used to find matrix $E$. It indicates that the Mex program is error prone.

$$C = \begin{bmatrix} 13 & 18 \\ 17 & 24 \\ 21 & 30 \end{bmatrix}, \quad D = \begin{bmatrix} 13 & 18 \\ 17 & 24 \\ 21 & 30 \end{bmatrix}, \quad E = \begin{bmatrix} 7 & 10 \\ 19 & 28 \end{bmatrix}.$$

As it is pointed out that, when the conditions are not fully considered, there might be errors in the example. For instance, the multiplicability of the matrices was not considered in the program. Modifications should be made to make it general-purpose.

### 8.1.4 Mex file programming and procedures

In the above presentations the fundamental information of Mex programming was introduced. With the information, the following procedures can be used to write C Mex programs:

(1) Function `mexFunction()` is the main interface of MATLAB and Mex program. The pointers of all the input arguments are passed to C code. The function name is fixed, and cannot be changed to other names.

(2) Header files should be included in the C code, such that function `mxGetPr()` can be used to retrieve the pointers of the input arguments. Functions `mxGetM()` and `mxGetN()` can be used to extract the row and column numbers of the input arguments. The variables can then be extracted from the MATLAB caller.

(3) The function `mxCreateDoubleMatrix()` can be used to allocate memory for a matrix. The pointer of the matrix can be retrieved with `mxGetPr()`. The returned variables to the caller can be assigned in C code.

(4) When the C programming is completed, command `mex` can be used to compile the code, and generate the executable file.

(5) A *.m file can also be written to provide the online help facilities. The Mex executables can be called in exactly the same way, as if it was just a MATLAB function. Examples are provided later to illustrate the Mex programming techniques.

**Example 8.2.** Consider the fractional tree problem in Example 4.18. Compare the efficiencies of MATLAB and Mex functions.

**Solutions.** It is known that large loops are involved in the example. An exaggerated loop of 10 000 000 points is used in the comparison. The following MATLAB statements can be used, and the total elapsed time, excluding the generation of the random vector, is 2.57 seconds.

```
>> N=10000000; v=rand(N,1); x=0; y=0; tic,
   for k=2:N, gam=v(k);
      if gam<0.05, x(k)=0; y(k)=0.5*y(k-1);
      elseif gam<0.45,
         x(k)=0.42*(x(k-1)-y(k-1)); y(k)=0.2+0.42*(x(k-1)+y(k-1));
      elseif gam<0.85,
         x(k)=0.42*(x(k-1)+y(k-1)); y(k)=0.2-0.42*(x(k-1)-y(k-1));
      else, x(k)=0.1*x(k-1); y(k)=0.1*y(k-1)+0.2;
   end, end; toc
```

It can be seen that the loops in MATLAB programming are not quite efficient. A Mex file can be tried with C code, in `frac_tree1.c`. Four input arguments and two returned arguments are designed for the Mex file. The first two input arguments are the initial points in $x_0$ and $y_0$. The third argument is a random vector, and the fourth one is the number of points. The returned arguments are vectors $x$ and $y$, where `mxCreateDoubleMatrix()` function is called in C code for their allocation. The listing of the C code is given below.

```
#include "mex.h"
void mexFunction(int nl, mxArray *pl[], int nr, const mxArray *pr[])
{ double *xp, *yp, *vp, *x0p, *y0p, *v0p, x0, y0;
  double vv; long i, N;
  x0p=mxGetPr(pr[0]); y0p=mxGetPr(pr[1]);
  v0p=mxGetPr(pr[2]); N=mxGetScalar(pr[3]);
  pl[0]=mxCreateDoubleMatrix(N,1, mxREAL);
  pl[1]=mxCreateDoubleMatrix(N,1, mxREAL);
  xp=mxGetPr(pl[0]); yp=mxGetPr(pl[1]);
  xp[0]=x0p[0]; yp[0]=y0p[0];
  for (i=1; i<N; i++){vv=v0p[i];
     if (vv<0.05){yp[i]=0.5*yp[i-1]; xp[i]=0;}
```

```
        else if (vv<0.45) {xp[i]=0.42*(xp[i-1]-yp[i-1]);
            yp[i]=0.2+0.42*(xp[i-1]+yp[i-1]);}
        else if (vv<0.85) {xp[i]=0.42*(xp[i-1]+yp[i-1]);
            yp[i]=0.2-0.42*(xp[i-1]-yp[i-1]);}
        else {xp[i]=0.1*xp[i-1]; yp[i]=0.1*yp[i-1]+0.2;}
}}
```

When the file is compiled, an executable file `frac_tree1.mexw64` can be generated. The total elapsed time for the same problem is reduced to 0.17.

```
>> N=10000000; v=rand(N,1);
   tic, [x1,y1]=frac_tree1(0,0,v,N); toc
```

It can be seen that when the Mex technique is adopted, the efficiency is significantly boosted. For this example, it is about 15 times faster, when C programming is involved. Therefore, for problems with intensive use of loops Mex programming can be adopted.

It can be seen from the previous examples that the data types of the input and returned variables are all `double()` matrices. In fact, in Mex programming, various other data types, such as multidimensional array, structured variables, cell arrays and strings, are all supported. The use of these data types is not as straightforward as in double precision data types. In the next section, different data types are demonstrated with examples, while in actual programming double matrices are recommended and should be used whenever necessary.

## 8.2 Mex manipulation of different data types

In this chapter, the interface between MATLAB and C is presented, and the main target is how to feed the arguments from MATLAB into a C program, or how the arguments in C can be transferred back to the caller in the form acceptable in MATLAB. Besides, data file transfer between MATLAB and C is also explored. Different data types and their conversions in C language are the internal problems in C and they are not discussed here.

Apart from the numerical argument transfer, the argument transfers of other data types are also supported in Mex programs, for instance, string variables, multidimensional arrays, and cells. Operations on MATLAB data files are also supported in Mex files. In this section, the transfer of a variety of data types is demonstrated in Mex files.

### 8.2.1 Processing of various input and output data types

In the Mex file, other types of data structures can be used, but their handling is not as simple as in double precision data type. In this section, various data types are demonstrated in Mex format. We shall introduce some commonly used C functions defined in `mex.h` file.

(1) Dynamic allocations. Similar to the dynamic allocation function `alloc()` in ANSI C, function `mxMalloc()` can be used in Mex format, to create a dynamic vector space with `void *mxCalloc(n,sizeof(variable type))`. The function can be used to create a vector space with $n$ elements. The size of the space can be adjusted automatically with the data types of the variables. The total space required can be set with function `sizeof()`. For instance, if one wants to create a double precision space with $n$ elements, function `mxCalloc(n,sizeof(double))` can be used.

(2) Free allocated spaces. Function `void mxFree(void *ptr)` can be used to implement this action. Function `void mxDestroyArray(mxArray *ptr)` can be used to free the dynamic space created by functions like `mxCreate`.

(3) Get variable name. The following command can be used to find the name of the $(k + 1)$th input argument:

```
const char *mxGetName(const mxArray *pr[k]);
```

(4) The output statements defined in Mex are similar to those in ANSI C. For instance, in `mex.h` file, function `mexPrintf()` is defined, which is similar to `printf()`, with the syntax `int mexPrintf(format,variable list)`.

### 8.2.2 Reading and writing of string variables

Mex programs can be used to manipulate strings. The string extraction statement in Mex is `mxGetString()`, and the output function is `mxCreateString()`. The following examples are used in the demonstration.

**Example 8.3.** Write a C Mex program, which can be used to accept input strings, and which has the following actions to respond:
(1) If there is no returned argument, the string results can be displayed in MATLAB command window.
(2) If there is one returned argument, the input string is returned directly.
(3) Error messages should be displayed for too many arguments.

**Solutions.** Based on the above requirements, the following C program can be written. The file name is `mex_string.c`, and the executable file `mex_string.mexw64` can be obtained.

```
#include "mex.h"
void mexFunction(int nl, mxArray *pl[], int nr,const mxArray *pr[])
{ char *mystr; int nstr, i;
    if (nr!=1) mexErrMsgTxt("Wrong number of input arguments!");
```

```
   if (nl>1) mexErrMsgTxt("Too many output arguments!");
   if (!mxIsChar(pr[0])) mexErrMsgTxt("String input is required!");
   nstr=mxGetNumberOfElements(pr[0])+1;
   mystr=mxCalloc(nstr, sizeof(char));
   mxGetString(pr[0],mystr,nstr);
   if (nl==1) pl[0]=mxCreateString(mystr);
   else {for (i=0;i<nstr;i++)
   mexPrintf("%c",mystr[i]); mexPrintf("\n");} mxFree(mystr);
}
```

where the two string-related functions are:
- Function mxGetString(ptr,mystr,nstr) can be used to assign the pointer ptr to the string mystr.
- Function pl[k]=mxCreateString(str) can be used to assign the string variable str to the $(k+1)$th returned argument.

When the following executable file is called

```
>> mex_string('This is an string.')
   S=mex_string(str2mat('This is an string','problem?'))
```

the first function displays "This is an string." whereas the string displayed by the second call is "Tphriosb liesm ?a n s t r i n g". The first string is the one we are expecting, while, in the second one, the characters appear to be disordered. It can be seen that the characters in the first and second strings are arranged in alternating orders, to form the final string. Therefore it can be seen that this C program cannot handle string matrices well. We can use alternative MATLAB commands to post-process the needed strings.

```
>> reshape(S,2,length(S)/2)
```

so that the correct two strings can be extracted. Besides, Mex file can also be modified to get directly the correct strings. For instance, in the C program below, the string matrices can be handled directly, with the function mxCreateCharArray(). Multidimensional string arrays can be established with such a function. Meanwhile, we can read out the sizes of the strings so that they can be returned directly, and function mxChar can be called to create string variables in Mex files.

```
#include "mex.h"
void mexFunction(int nl,mxArray *pl[],int nr,const mxArray *pr[])
{ char *mystr; mxChar *xp; int nstr, i, j, *ndims, m, n;
  nstr=mxGetNumberOfElements(pr[0]);
```

```
mystr=mxCalloc(nstr, sizeof(char));
ndims=mxGetDimensions(pr[0]); m=ndims[0];
n=mxGetNumberOfDimensions(pr[0]);
mxGetString(pr[0], mystr, nstr);
if (nl==1) {pl[0]=mxCreateCharArray(n,ndims);
xp=(mxChar *) mxGetData(pl[0]);
for (i=0; i<nstr; i++) xp[i]=mystr[i];}
else {for (j=0; j<m; j++){
for (i=j;i<nstr;i+=m)mexPrintf("%c",mystr[i]); mexPrintf("\n");}}
mxFree(mystr);
}
```

The executable file `mex_strmat.mexw64` can be generated after compilation, and when it is executed, correct results are obtained.

```
>> mex_strmat('This is an string.')
   S=mex_strmat(str2mat('This is an string','problem?'))
```

It can be seen that this function can be used to process string vectors as well as multidimensional string arrays. An example will be given later to demonstrate the processing of the expected data types.

### 8.2.3 Processing of multidimensional arrays

Multidimensional array arguments are also supported in C Mex files. The function `mxGetPr()` can still be used to handle multidimensional arrays. The creation function for the array is `mxCreateNumericArray()`, with the syntax

$p$=mxCreateNumericArray($n$,$m$)

where $p$ is the pointer of the multidimensional array, $n$ is the number of dimensions, and $m$ is a vector storing the sizes in each dimension. An example will be given later to demonstrate the input, output, and manipulation of multidimensional arrays.

**Example 8.4.** The aim of the example is to write a C program in Mex format. For the first and second dimensions of a multidimensional array, transpose can be made, and the new variable can be returned.

**Solutions.** Based on the request, the following C program can be written. The file name is `mex_mattrans.c`. After the compilation, the executable file `mex_mattrans.mexw64` can be created.

```
#include "mex.h"
void mexFunction(int nl,mxArray *pl[],int nr,const mxArray *pr[])
```

```
{ double *Ap,*Bp;
  int mA,nA, m_total, n_elements, *ndims, n, *n1, i, j, k;
  n_elements=mxGetNumberOfElements(pr[0])+1;
  n=mxGetNumberOfDimensions(pr[0]);
  ndims=mxGetDimensions(pr[0]);
  Ap=mxGetPr(pr[0]); n1=mxCalloc(n,sizeof(int));
  mA=ndims[0]; nA=ndims[1]; n1[0]=nA; n1[1]=mA;
  for (j=2; j<n; j++) n1[j]=ndims[j];
  pl[0]=mxCreateNumericArray(n, n1, mxDOUBLE_CLASS, mxREAL);
  Bp=mxGetPr(pl[0]); m_total=nA*mA;
  for (k=0; k<n_elements; k+=m_total) for (i=0; i<mA; i++)
  for (j=0; j<nA; j++) Bp[k+i*nA+j]=Ap[k+j*mA+i]; mxFree(n1);
}
```

The following commands can be used to generate a three-dimensional array for testing:

```
>> A0=[1 3 5 2 4 6 7 9 11 8 10 12 13 15 17,...
        14 16 18 19 21 23 20 22 24];
   A=reshape(A0,3,2,4); size(A)
```

Calling mex_mattrans.mexw64 function, a new three-dimensional array can be returned. To see whether it is the same as the one expected, it can be seen from the third page in the third dimension that the required three-dimensional array is indeed generated.

```
>> B=mex_mattrans(A); size(B)
```

The results obtained are

$$A(:,:,3) = \begin{bmatrix} 13 & 14 \\ 15 & 16 \\ 17 & 18 \end{bmatrix}, \quad B(:,:,3) = \begin{bmatrix} 13 & 15 & 17 \\ 14 & 16 & 18 \end{bmatrix}.$$

### 8.2.4 Processing of cells

Cell arrays are also supported in Mex. The $p$=mxCreateCellMatrix($m,n$) function can be used to generate an $m \times n$ cell array at pointer $p$. The cell array can also be used as a one-dimensional array in C, which is the same as the matrices in C. Commands mxGetM() and mxGetN() can also be used to retrieve the row and column numbers of a cell array. If necessary for handling multidimensional cell arrays, function mxCreateCellArray() can also be called to create multidimensional cell arrays, and the syntax is the same as that for multidimensional arrays.

If the contents in the $i$th cell is expected, function mxGetCell() can be used to get its pointer. The syntax of the function will be demonstrated through the following examples. If one wants to write information to a cell, function mxSetCell() can be called.

**Example 8.5.** Write a simple C program to make the "transpose" of the $n \times m$ input cell array, and return the result back to the caller.

**Solutions.** Based on the request, the following Mex C program can be written. The function is stored in the file mex_cell.c. The sizes $n$ and $m$ of the input cell array are measured, and then a loop can be used to process each cell, that is, to read out the contents in the $(i, j)$th cell with function mxGetCell(), and then use mxSetCell() function to write it to the $(j, i)$th cell of the returned argument, to implement the "transpose".

```
#include "matrix.h"
void mexFunction(int nl,mxArray *pl[],int nr,const mxArray *pr[])
{ mxArray *xp; int i, j, m, n;
  m=mxGetM(pr[0]); n=mxGetN(pr[0]);
  pl[0]=mxCreateCellMatrix(n,m);
  for (i=0; i<m; i++) {for (j=0; j<n; j++) {
    xp=mxGetCell(pr[0],j*m+i); mxSetCell(pl[0],i*n+j,xp);
}}}
```

For instance, the following cell array can be constructed:

```
>> C={1, [1 2 3;4 5 6]; 2, 'Test Cell Matrix';
       [1 2; 3 4], [3,2,1; 4,2,0]}
```

With the call to the executable file mex_cell.mexw64, the following results are obtained. It can be seen that the results are as expected.

```
>> D=mex_cell(C)
```

It can be seen from the above example that the arguments of these types cannot be easily transferred in C programs. Apart from the above few data types, the definition of the commonly used structured variables is even more complicated. They are not demonstrated here. It is suggested that if there is no specific reason, the normal matrix form should be used in data transfer.

### 8.2.5 Reading and writing of MAT files

With complicated programming in MATLAB, sometimes, programs written in other languages are often used. Theoretically speaking, executable files written in any other language can be called directly from MATLAB functions. In earlier versions, the syntaxes were simple, and the procedures are:
(1) Pass the information to the executable file into a data file.
(2) Rewrite the source of the executable file to read data from the data file.
(3) Run the executable file by adding exclamation mark (!) to the file name. The executable file should be recompiled and linked again.
(4) Write the results back to the data file. After the execution process, the data file can be read from MATLAB commands. Now a simple example is used to show the syntaxes.

It can be seen that under such a calling syntax, the key issue is how to use other languages to read and write MAT files. In this section, C language is used as an example to demonstrate the information exchange between MATLAB and C. In fact, there are interfaces between MATLAB and Fortran.

The following functions can be used to access MATLAB files:
– Open and close MAT files. Generally, MATLAB commands save and load can be used to access MAT files. C language functions matOpen() and matClose() can be used to open and close MAT files, with the syntaxes:

```
MATFile *matOpen(file name string, file type string);
int matClose(MATFile *mfp);
```

where the data type of file name handle is MATFile. Two input arguments are used in the function, the former is the file name, while the latter is the type of the string, for instance, "r" and "w". The returned argument in the first function is the MAT file handle. The input argument in the second function is also the handle mfp. Note that, since C programming is used here, the string should be defined in double quotation marks. The two functions may return information. In the first function, if the returned argument is not NULL, the file opening process is successful, and NULL is returned after an unsuccessful attempt. In the second function, the returned argument 0 means the closing process is successful, 1 is returned for an unsuccessful try.
– Read MAT files. The commonly used function is

```
mxArray *matGetNextArray(MATFile *mfp);
```

The following functions can be used to read the arrays stored in the MAT file one by one:

- Write MAT files. If one wants to write a variable into a MAT file, the variable must be named, so that it can be loaded with `load` command. The function to call in C is

```
int void mxSetName(mxArray *ptr, const char *name);
```

where `name` is the variable name to assign. With the variable name setting, function `matPutArray()` can be used to store a variable pointed by `ptr` into the MAT file.

```
int matPutArray(MATFile *mfp, const mxArray *ptr);
```

- Variable management in MATLAB files. The following functions can be used to get the folder and file names:

```
char **matGetDir(MATFile *mfp, int *num);
FILE *matGetFp(MATFile *mfp);
```

The former is used to get variable name in the MAT file, while the latter is used to get the pointer of the MAT file in C program.

**Example 8.6.** An example is used here to show how to use the primitive method to call C programs. Of course, the characteristics of MATLAB are not well displayed, since the problem can be better implemented in MATLAB in a simple and straightforward way, and faster.

**Solutions.** Assume that there are two matrices $A$ and $B$. We want to compute their product $C=A*B$ in C language. The following standalone C program can be written:

```
#include "mat.h"
main(int argc, char **argv)
{ MATFile *fp1,*fp2; int i,j,k, mA,nA, mB, nB, *ndims;
  mxArray *pA, *pB, *pC; double *A, *B, *C;
  fp1=matOpen(argv[1],"r");
  if (fp1 == NULL) {
     printf("Error opening file %s\n",argv[1]); return(1);}
  pA=matGetNextArray(fp1); pB=matGetNextArray(fp1);
  ndims=mxGetDimensions(pA); mA=ndims[0]; nA=ndims[1];
  ndims=mxGetDimensions(pB); mB=ndims[0]; nB=ndims[1];
  if (nA!=mB) {
     printf("Matrices A, B not compatible for multiplication\n");
     return(1);}
  A=mxGetPr(pA); B=mxGetPr(pB);
  pC=mxCreateDoubleMatrix(mA,nB,mxREAL); C=mxGetPr(pC);
  for (i=0; i<mA; i++){for (j=0; j<nB; j++){C[i*mA+j]=0;
```

```
    for (k=0; k<nA; k++) C[i*mA+j]+=A[k*mA+i]*B[j*nA+k];}}
    matClose(fp1); fp2=matOpen(argv[2],"w");
    mxSetName(pC, "C"); matPutArray(fp2,pC); matClose(fp2);
}
```

It can be seen that the C program is a completed one. It is not necessary to use Mex file to set up the interface with MATLAB. The syntax of the function is

`mex_matp` file name 1  file name 2

If the files cannot be found, the error message will be displayed. If the file exists, the two matrices can be loaded. They are validated and checked to see whether they are compatible for multiplication. If they are not compatible, error messages are displayed. If they are compatible, the two matrices can be extracted and then multiplied. The results are stored in the output file, with a variable name of **C**. The following MATLAB commands are used:

```
>> A=[1 2 3; 4 5 6; 7 8 0]; B=[1 2 3 4; 5 6 7 8; 9 10 11 12];
   save mat_tmp1 A B, !mex_matp mat_tmp1.mat mat_tmp2.mat
   load mat_tmp2; C
```

In fact, the program can also be executed under DOS environment. No exclamation mark is needed to run the program. After the execution of the program, the same `mat_tmp2.mat` file can be generated. To read the results, MATLAB environment should still be used. Therefore the program cannot be regarded as a completely standalone program to MATLAB. In actual programming, it is recommended that the variables are transferred with pointers, rather than using data files. The transfer of large-scale matrices is rather slow, and the syntaxes are not regularized. More MAT files are needed.

## 8.3 Direct calling of MATLAB functions from C programs

The benefit of using C programming is that it is fast. Since the integrity of C language programs is rather low, it is rather difficult to write C programs for numerical computation tasks. Therefore the highly integrated MATLAB functions may be needed. The interface between MATLAB and C makes this possible. The user may call directly MATLAB functions within the C programs. Since the DLL file thus generated can only be used within MATLAB, the C program can only be called from MATLAB environment. In C programs, MATLAB functions can be called with

```
int mexCallMATLAB(int nlhs, mxArray *plhs[], int nrhs,
      mxArray *prhs[], const char *command_name);
```

In fact, the syntax here is rather close with the entry function in the Mex format. At the end, an extra argument `command_name` is used, which is a string of MATLAB function name. In the format, arguments like nlhs are used to show the differences with the Mex entry functions.

**Example 8.7.** Call the test.m function designed in Example 5.1 in the Mex format.

**Solutions.** The file needed to call is a script file, so there are no input and output arguments. The arguments nlhs and nrhs should be set to 0's. The two pointers should be set to NULL. The following C program can be written:

```
#include "matrix.h"
void mexFunction(int nl,mxArray *pl[],int nr,const mxArray *pr[])
{ mxArray *xp[1];
  if (nr!=0) mexErrMsgTxt("Wrong number of input arguments!");
  if (nl>0) mexErrMsgTxt("Too many output arguments!");
  mexCallMATLAB(0,NULL,0,NULL,"test");
}
```

The file mex_test.mexw64 can be created by the compiling process, and the execution is the same as the test command.

**Example 8.8.** Write a C program in the Mex format, such that its mexw64 file can be called in the syntax:

$[B,D,V]$=mex_eigens$(A,n)$

where matrix $A$ is given, $n$ is the power of $A$, and the returned $B$ is $A^n$. The other two returned arguments $D$ and $V$ are respectively the eigenvalue and eigenvector matrices of $B$.

**Solutions.** The tasks listed above can easily be implemented in MATLAB as

```
>> A=[1 2 3; 4 5 6; 7 8 0]; n=3; b=A^3
   [v,e]=eig(b); d=diag(e), v
```

The pure MATLAB function can be written as

```
function [B,D,V]=mex_eigens(A,n)
B=A^n; [V,d]=eig(A); D=diag(d);
```

The example here is to demonstrate how to use C language to call MATLAB functions. We concentrate on how to set pointers and call MATLAB functions. Based on the requests of the example, the tasks are decomposed as follows:
(1) Compute $A^n$ and assign it to $B$. The corresponding MATLAB function to compute the power is mpower(), with the syntax $B$=mpower$(A,n)$. In the function call of mpower(), two input and one output arguments are used.

(2) When **B** is obtained, function [**V**,**d**]=eig(**B**) can be used to compute the eigen-value matrix **D** and eigenvector matrix **V**. The function has one input and two returned arguments.

(3) With the facilities in C language, the diagonal elements in matrix **d** can be extracted and assigned to vector **D**, and finally, the code returns the two arguments **D** and **V**.

Based on the above explanations, the following C program in the Mex format can be written:

```
#include "matrix.h"
void mexFunction(int nl,mxArray *pl[],int nr,const mxArray *pr[])
{ mxArray *xp[1], *L[2]; int n,m, i;
  double *A, *pA, *pB1, *pB2, *pB3, *pA1;
  m=mxGetM(pr[0]); n=mxGetN(pr[0]);
  mexCallMATLAB(1,xp,2,pr,"mpower"); pA1=mxGetPr(xp[0]);
  pl[0]=mxCreateDoubleMatrix(m,n,mxREAL); pB1=mxGetPr(pl[0]);
  pl[1]=mxCreateDoubleMatrix(m,1,mxREAL); pB2=mxGetPr(pl[1]);
  pl[2]=mxCreateDoubleMatrix(m,n,mxREAL); pB3=mxGetPr(pl[2]);
  for (i=0; i<m*n; i++) pB1[i]=pA1[i];
  mexCallMATLAB(2,L,1,xp,"eig");
  pA1=mxGetPr(L[0]); for (i=0; i<m*n; i++) pB3[i]=pA1[i];
  pA1=mxGetPr(L[1]); for (i=0; i<m; i++) pB2[i]=pA1[i+i*m];
}
```

With such a C program, the Mex executable file mex_eigens.mexw64 can be generated. Then the following commands can be used to compute the matrices using two methods:

```
>> A=[1 2 3; 4 5 6; 7 8 0]; n=3; [B,D,V]=mex_eigens(A,n)
   b=A^n; [v,d]=eig(b); d=d(:); [norm(B-b), norm(D-d), norm(V-v)]
```

When called from MATLAB, the following results are returned:

$$B = \begin{bmatrix} 279 & 360 & 306 \\ 684 & 873 & 684 \\ 738 & 900 & 441 \end{bmatrix}, \quad D = 1000 \begin{bmatrix} 1.7816 \\ -0.0001 \\ -0.1886 \end{bmatrix},$$

$$V = \begin{bmatrix} 0.2998 & 0.7471 & -0.2763 \\ 0.7075 & -0.6582 & -0.3884 \\ 0.6400 & 0.0931 & 0.8791 \end{bmatrix}.$$

It can be seen that the result of the Mex format program is exactly the same as that in MATLAB. C programs in the Mex format can also be written to open MATLAB graphics windows and draw plots. Handle graphics command can also be used to modify properties of the plots. Later, examples will be given to show how to modify graphical object properties.

**Example 8.9.** Draw sine and cosine curves using a C program in the Mex format.

**Solutions.** Sine and cosine functions can be drawn with MATLAB function plot(), which can be called directly within C language, and an argument is returned in the function. It is the handle of the curves. The following C program can be written:

```
#include "matrix.h"
#define M 63
void mexFunction(int nl,mxArray *pl[],int nr,const mxArray *pr[])
{ mxArray *R[2]; int i; double *pB1, *pB2, t;
    if (nr!=0) mexErrMsgTxt("Wrong number of input arguments!");
    if (nl>1) mexErrMsgTxt("Too many output arguments!");
    R[0]=mxCreateDoubleMatrix(1,M,mxREAL); pB1=mxGetPr(R[0]);
    R[1]=mxCreateDoubleMatrix(2,M,mxREAL); pB2=mxGetPr(R[1]);
    t=0;
    for (i=0; i<M; i++) {
        pB1[i]=t; pB2[2*i]=sin(t); pB2[2*i+1]=cos(t); t+=0.1;}
    mexCallMATLAB(1,pl,2,R,"plot");
}
```

If in MATLAB environment, program mex_mysin1.mexw64 is executed

```
>> h=mex_mysin1
```

the returned handles are $h = [1.0024, 74.0022]^T$. A graphics window is opened automatically. The sine and cosine curves can be drawn in the window. The result is the same as when calling plot() function in MATLAB. Besides, the two handles are returned in argument h.

When writing C programs, one must bear in mind that:

(1) Although numerical computation is involved in the program, the header file math.h is not needed to be included again, since it has already been included in the header file matrix.h. Of course, it does not matter if it is included again.

(2) Before compiling the program to generate the mexw64 file, clear command should be used to remove the function from MATLAB workspace, otherwise it may be incorrectly compiled. It is better cleared, if there is no graphical facilities involved in the C program, otherwise, although compiling can be made, the most recent version may not be called, since there is already a compiled version in MATLAB workspace. Two functions mexGet() and mexSet() are also provided in Mex.

Their actions are similar to the MATLAB functions get() and set(), used to get and set properties for the given handle. The syntaxes of the two functions are

```
const mxArray *mexGet(double handle, const char *property);
int mexSet(double handle,const char *property,mxArray *value);
```

where the handle names are exactly the same as those in MATLAB. Note that double quotation marks should be used in C instead. Besides, in C programming, pointers should be used. In the latter function, if 0 is returned, successful set action is taken, otherwise the returned argument is 1.

**Example 8.10.** Consider again the plotting program in Example 8.9. If the thickness of the sine curve is expected to be increased by 10 times, rewrite the program.

**Solutions.** We may consider increasing the thickness of the first curve by 10 times, and write a C program in Mex format. The file name is mx_plot.c.

```
#include "matrix.h"
void mexFunction(int nl,mxArray *pl[],int nr,const mxArray *pr[])
{ mxArray *R[2], *L[1], *lWidth; int i, M; M=63;
  double *pB1, *pB2, t, *pH, *lw;
  if (nr!=0) mexErrMsgTxt("Wrong number of input arguments!");
  if (nl>0) mexErrMsgTxt("Too many output arguments!");
  R[0]=mxCreateDoubleMatrix(1,M,mxREAL); pB1=mxGetPr(R[0]);
  R[1]=mxCreateDoubleMatrix(2,M,mxREAL); pB2=mxGetPr(R[1]);
  t=0;
  for (i=0; i<M; i++) {
      pB1[i]=t; pB2[2*i]=sin(t); pB2[2*i+1]=cos(t); t+=0.1;}
  mexCallMATLAB(1,L,2,R,"plot");
  pH=mxGetPr(L[0]); lWidth=mexGet(pH[0],"LineWidth");
  lw=mxGetPr(lWidth); lw[0]=10*lw[0];
  mexSet(pH[0],"LineWidth",lWidth);
}
```

It can be seen that the programming is straightforward, while with MATLAB, a few commands are sufficient to implement all these:

```
>> t=0:.1:2*pi; h=plot(t,sin(t),t,cos(t));
   set(h(1),'LineWidth',10*get(h(1),'LineWidth'));
```

Therefore it is advised that, in scientific computation and computer graphics, if there are no special reasons, MATLAB programming is recommended, rather than C or other languages, not even calling MATLAB from C. If a certain task must be programmed with C, Mex format is recommended to exchange directly information with MATLAB, otherwise the chance of error or trouble may be increased.

## 8.4 Standalone program conversion from MATLAB functions

It can be seen in the previous presentation that when C is used to call MATLAB functions, the obtained executable files are not standalone files. They cannot be executed without running MATLAB. In fact, some functions in MATLAB can be automatically translated into C programs, and with certain compilers, standalone programs can be generated, which can be executed without MATLAB.

**Example 8.11.** Consider again the fractional tree problem in Example 8.2. Write an independent executable in C language.

**Solutions.** In fact, there is no need to rewrite the MATLAB function in C language. The original MATLAB script can be rewritten to a MATLAB function first. For instance, add a statement beginning with function, then, with input() function allow the user to input the value of $N$. With the number $N$, the fractional tree can be computed and drawn. The modified MATLAB function is

```
function c8e_tree
N=input('Input the number of points N    ');
v=rand(N,1); x=0; y=0;
for k=2:N, gam=v(k);
   if gam<0.05, x(k)=0; y(k)=0.5*y(k-1);
   elseif gam<0.45,
      x(k)=0.42*(x(k-1)-y(k-1)); y(k)=0.2+0.42*(x(k-1)+y(k-1));
   elseif gam<0.85,
      x(k)=0.42*(x(k-1)+y(k-1)); y(k)=0.2-0.42*(x(k-1)-y(k-1));
   else, x(k)=0.1*x(k-1); y(k)=0.1*y(k-1)+0.2;
end, end
plot(x,y,'.')
```

With such MATLAB function, the command mcc -m c8e_tree can be used to convert the file automatically into an executable one, c8e_tree.exe. The file can be executed directly, without the support from MATLAB. It is worth mentioning that the file generated is an exe file. There is no C source file returned.

**Example 8.12.** Obtain the C source code for the program in Example 8.11.

**Solutions.** If command mcc -l c8e_tree is used, a set of complete C language programs are generated, including *.c, *.h files, and so on. The generated C source file is rather lengthy, and the kernel part of the program is hidden in c8e_tree.dll file, the source is, in fact, not visible. The source code is not listed here. It should be said that the automatically translated C source code is not a genuine C code. Therefore, if it is not really necessary, do not use this programming format in scientific programming.

A method for generating C files is given in Example 8.11. The advantage is that the generated standalone file can be executed independently in operating systems such as Microsoft Windows, without the support of MATLAB. There are limitations in the executable files, since the commonly used eval() function is not supported. Also the facilities such as anonymous function, symbolic computation, and Simulink are not supported. The input and returned arguments are not supported. Therefore the programming style is not at all practical, and it is not recommended to using this programming format.

## 8.5 Problems

8.1 Write a Mex program such that it may accept an arbitrary number of input matrices. Take transposes of these matrices and return the results to the caller, as returned arguments.

8.2 Refer to the Mex format, and modify in C language the function in Example 7.37. Convert the Brownian motion animation program in MATLAB file to a function, and convert it into a standalone executable file.

8.3 In Chapter 10, the design of a graphical user interface will be introduced, with design examples. Executable files can be generated from these examples. Convert the "Hello World" interface c10exgui1() and multimedia player function c10mmplay() into executable files, and run these files. Can you convert the three-dimensional graph drawing interface c10eggui3() in Example 10.12 into an executable file? Why? Validate your prediction with the example.

# 9 Fundamentals in object-oriented programming

Many programming skills were discussed earlier in the previous chapters. The programming styles discussed so far can be regarded as regular programming methods. In this chapter and Chapter 10, a brand new programming method – object-oriented programming (OOP) – is presented. In this chapter, the design and usage of classes are presented, while in Chapter 10, graphical user interface design is addressed.

In Section 9.1, concepts and fundamental knowledge of object-oriented programming are introduced. The necessity of object-oriented programming is discussed, and the difference between this and regular programming methods is addressed. In Section 9.2, the design of classes is presented, based on an example of a special pseudo-polynomial. In Section 9.3, algebraic operations are carried out, and function implementation is elaborated. The pseudo-polynomial is still used as an example to implement addition, subtraction, multiplication and power of the object. In Section 9.4, based on the pseudo-polynomial class, a higher-level fractional-order transfer function class is introduced. This new class can be designed on top of an existing class.

## 9.1 Concepts of object oriented programming

The object-oriented programming technique is an important computer programming method. This programming mechanism is well supported in MATLAB. In this section, the fundamental concepts and knowledge of object-oriented programming is introduced, and the concepts of class and object are proposed. They form the systematical foundation of object-oriented programming techniques.

### 9.1.1 Classes and objects

Compared with tradition programming approaches, object-oriented programming is a brand new framework. Some fundamental concepts of object-oriented programming are presented first in this section.

**Definition 9.1.** Object-oriented programming is a specific programming format. The concept of objects is adopted, and based on it programs can be written.

**Definition 9.2.** The object is a representation method of data. The data are represented as fields in the objects. The fields are also known as attributes, or member functions.

**Definition 9.3.** A collection of objects having the same structure and behavior is referred to as a class.

https://doi.org/10.1515/9783110666953-009

**Definition 9.4.** The processing functions in object-oriented programming techniques are also known as the methods of objects.

If a class is designed, an object can be regarded as an instance of the class. All the fields and methods of the class can be used by the object.

In object-oriented programming applications, the fields can be retrieved and modified with the relevant methods, i. e., setting the attributes of the object.

There are significant differences in the object-oriented programming and regular programming mechanisms. In regular programming, the statements are executed statement by statement, while in object-oriented programming, objects and methods are defined. The functions are normally idle. If an event happens, the object is triggered to execute the appropriate responding functions. Microsoft Windows is an example of object-oriented programming. If one clicks a menu item, an event happens, and Windows mechanism executes automatically the corresponding code to react to the menu selection event. The object-oriented programming applications are everywhere, and this importance of such programming is evident.

Compared with the regular programming mechanism, the readability, reusability, and extendability of the functions in object-oriented programming is much superior.

For ordinary readers, we'll mention that two different kinds of programming technique are used in object-oriented programming. One is called the client method, while the other is the programmer's method[15]. In the former format, the reader can use the existing classes and objects directly in MATLAB, there is no need to bother with the low-level programming details. Most readers are users of object-oriented programming techniques. In the programmer's format, the user should learn low-level programming details, and be able to create classes and write low-level code.

In this chapter, the full process of object-oriented programming is demonstrated with a simple example. If the low-level programming skills are learnt by the users, they may have better capabilities in using MATLAB to solve scientific computing problems directly.

## 9.1.2 Data type of classes and objects

A variable name can be assigned in MATLAB to represent a class. For instance, the tf class is provided in Control System Toolbox to represent a class of transfer functions. Also, class ss is also provided to define the class of state space models. When the class is defined, a variable name can be used to describe an object of the class. For instance, $G$ can be used as a variable name of a transfer function object.

Many fields are usually defined under a class. For instance, in the transfer function class, the fields of the numerator and denominator coefficients are needed, which can be expressed as field names num and den. Extraction or modification of the fields is usually needed in object-oriented programming. Extraction and assignment of the fields should be made. If Bode diagram of the object is needed, a relevant method is required. For instance, function bode(G) in Control System Toolbox can be used to draw Bode diagrams directly, and function bode() is one of the methods of the object.

**Example 9.1.** The transfer function of a control system is given below. Input the model into MATLAB workspace, and find the fields of the object

$$G(s) = \frac{s + 8}{s(s^2 + 0.2s + 4)(s + 1)(s + 3)}.$$

**Solutions.** Client point of view is used here to understand the object-oriented programming applications in this example. Various transfer function input methods are provided in the Control System Toolbox. For instance, a Laplace operator $s$ can be declared first, then simple expression can be used to input the model into MATLAB, and finally, the result can be displayed directly.

```
>> s=tf('s'); G=(s+8)/s/(s^2+0.2*s+4)/(s+1)/(s+3)
```

With the MATLAB command get(G), the fields of the class tf can be displayed as follows:

```
>> get(G)

    Numerator: {[0 0 0 0 1 8]}
  Denominator: {[1 4.20000 7.800000 16.600000000000001 12 0]}
     Variable: 's'
      IODelay: 0
   InputDelay: [0×1 double]
  OutputDelay: [0×1 double]
           Ts: 0
        . . .
```

It can be seen that if a class is expected, the name of the class should be chosen first, and a dedicated folder can be created for it. The MATLAB functions for the class should be placed inside the folder, with the class definition function, class display function, and the necessary overload functions. These topics are provided in the subsequent sections.

## 9.2 Design of classes

Class design is useful in scientific computing. In many of the toolboxes, the classes and objects are directly involved. Therefore, mastering the skills in the design and manipulation of classes may increase the users ability in better and creatively solving scientific computing problems.

### 9.2.1 The design of a class

The definition of a pseudo-polynomial is given below. The design of the class in MAT-LAB is demonstrated with a pseudo-polynomial. The definition is given first, the examples are used to illustrate the design procedures of classes.

**Definition 9.5.** The mathematical expression of a pseudo-polynomial is

$$p(s) = a_1 s^{\alpha_1} + a_2 s^{\alpha_2} + \cdots + a_n s^{\alpha_n}, \tag{9.2.1}$$

where $a_i$ is the coefficient and $\alpha_i$ is the order, $i = 1, 2, \ldots, n$. The orders here are not restricted to integers. Therefore, the polynomial here is referred to as a pseudo-polynomial.

To design a class, the following procedures can be taken:
(1) **Select a name for the class.** The regulations in the selection of class names are exactly the same as those in selecting variables names.
(2) **Create a blank folder.** The folder name should be the class name, proceeded by @. If the path of the current folder is under MATLAB search path, the new folder should not be added to the search path.
(3) **Design fields for the class.** Fields are used to store important parameters.
(4) **Two essential functions are needed.** To construct a class, there should be at least two MATLAB functions, one is the same as the name of the class, and used to allow users to input the class. The other is the `display.m` function, used to display the class.
(5) **Design necessary overload functions.** Any action of the class, including addition and multiplication, should be described by MATLAB functions dedicated to the class. It is better to assign the function names different from regularly used ones. For instance, to define addition action, the file name should be selected as `plus.m`, and the operator "+" can then be used to the class. The newly designed class comes with no methods. All the operations on the class should be described by the user-designed methods in MATLAB.

**Example 9.2.** Design a MATLAB class for the pseudo-polynomials.

**Solutions.** To design a class, its name should be selected by the user. For instance, the name ppoly could be selected. Then a blank folder @ppoly should be created in MATLAB search path. To describe uniquely the pseudo-polynomial, two vectors, $\boldsymbol{a} = [a_1, a_2, \ldots, a_n]$ and $\boldsymbol{\alpha} = [\alpha_1, \alpha_2, \ldots, \alpha_n]$, should be defined as the fields for the class ppoly, and the names can be selected as a and na, respectively.

### 9.2.2 Design and input of classes

The structure of class defining functions is fixed. It is slightly different from the standard MATLAB functions, however, it is easy to understand the structure. Examples are used to demonstrate the class defining functions.

**Example 9.3.** Write a class defining function for class ppoly.

**Solutions.** Before programming the function, the input format for the class should be considered carefully. It is fine if the syntaxes cannot fully be defined, since it can be extended to accept other syntaxes at any later time. For a pseudo-polynomial, the following three syntaxes are allowed:

$p$=ppoly($a,\alpha$), $p$=ppoly($a$), $p$=ppoly('s')

where, in the first one, the coefficient and order vectors are provided, while in the second, a vector of the coefficients is provided, and a normal polynomial can be constructed. In the third syntax, $p$ is declared as an $s$ operator. With these considerations, the MATLAB function can be written below.

```
classdef ppoly
    properties, a, na, end
    methods
        function p=ppoly(a,na)
        if nargin==1,
            if isa(a,'double'), p=ppoly(a,length(a)-1:-1:0);
            elseif isa(a,'ppoly'), p=a;
            elseif a=='s', p=ppoly(1,1); end
        elseif length(a)==length(na), p.a=a; p.na=na;
        else, error('Error: missmatch in a and na'); end
end, end, end
```

The function definition should not only give the input syntaxes, but also set the conversion methods. For instance, if only one vector is provided, the original pseudo-polynomial can be considered as having integer order, and the order vector should be defined as the integer order vector. Therefore, a ppoly object can be established. If the input argument is already a ppoly object, it can be passed to the returned argument directly.

With the `classdef` command used here, the functions like `subsasgn()` and `subsref()` are no longer necessary. Apart from the essential functions such as `ppoly.m` and `display.m`, the functions `get.m` and `set.m` should also be written, to complete the class defining process.

In the programming process, if a MATLAB method function is modified, the residential information of the function is not changed accordingly. Therefore, when a method is modified, the command `clear classes` should be issued to clear the residential information, and then the modified method can be called.

### 9.2.3 Class display

If a class is defined, a display function should be written, with the name `display()`. The function is called automatically whenever necessary. If an object is entered, and there is no semicolon to suppress its output, the object will be displayed automatically, by calling the `display()` function. The response actions should be programmed such that the object can be displayed in the desired format.

**Example 9.4.** Write a display function for the ppoly object.

**Solutions.** Considering Definition 9.2.1, a string should be written to display the pseudo-polynomial expression. Besides, certain simplification actions should be programmed. For instance, if there is the notation +1∗*s* in the string, 1∗ should be removed from the string. String substitution function `strrep()` should be used to substitute +1∗*s* with +*s* in the string. Write several examples of the strings and see what else should be substituted. In the following functions, some of the substitutions are made. If more substitutions are expected, just implement them in a similar manner. The final display function can be written as

```
function str=display(p)
np=p.na; p=p.a; if length(np)==0, p=0; np=0; end
P=''; [np,ii]=sort(np,'descend'); p=p(ii);
for i=1:length(p),
    P=[P,'+',num2str(p(i)),'*s^{',num2str(np(i)),'}'];
end
P=P(2:end); P=strrep(P,'s^{0}',''); P=strrep(P,'+-','-');
P=strrep(P,'^{1}',''); P=strrep(P,'+1*s','+s');
P=strrep(P,'*+','+'); P=strrep(P,'*-','-');
strP=strrep(P,'-1*s','-s'); nP=length(strP);
if nP>=3 & strP(1:3)=='1*s', strP=strP(3:end); end
if strP(end)=='*', strP(end)=''; end,
if nargout==0, disp(strP), else, str=strP; end
```

For ease of the subsequent programming, an extra argument `str` is reserved and allowed, and it can be used later.

**Example 9.5.** Input and display the pseudo-polynomial $p(s) = 3s^{0.7} + 4s + 5$.

**Solutions.** For the given pseudo-polynomial, the coefficient vector $\boldsymbol{a} = [3, 4, 5]$ and order vector $\boldsymbol{n} = [0.7, 1, 0]$ can be extracted. There is no need to change the sequences in the order vector. When the object is entered, the display() function can be called automatically, and the displayed string is 4*s+3*s^{0.7}+5. The pseudo-polynomial can be displayed in the descending order of $s$, while the collection of the like terms cannot be made automatically.

```
>> a=[3,4,5]; n=[0.7,1,0]; p=ppoly(a,n)
```

## 9.3 Programming of overload functions

The computation of objects depends heavily on the computing functions. In object-oriented programming, the computing functions can be regarded as the methods. In order to have the manipulations of the objects similar to those in regular MATLAB manipulation, it is better to select the same function names with other objects. If the function names are the same for different objects, the function is referred to as an overload function. Since overload functions are kept within the folder of the object, they may not conflict with other overload functions with the same name.

Algebraic computation mainly refers to addition, subtraction, multiplication, division, and exponentiation. The fundamental algebraic operations of ppoly objects are presented first, then overload function programming is illustrated.

**Theorem 9.1.** *If $p_1$ and $p_2$ are ppoly objects, then $p_1 + p_2$, $p_1 - p_2$, and $p_1 \times p_2$ are all ppoly objects, while $p_1/p_2$ is not a ppoly object.*

**Theorem 9.2.** *If $p$ is a decent ppoly object and $n$ is an integer, then $p^n$ is also a ppoly object, otherwise it is not. As a special case, if $p = s$, then $p^n$ is a ppoly object for any real number $n$.*

### 9.3.1 Overload addition functions

For a newly designed class, the operators such as +, −, and * cannot be used directly, MATLAB functions for the operators should be defined first. For the operators +, −, *, and ^, the function names should not be selected arbitrarily, corresponding function names must be selected as plus(), minus(), mtimes(), and mpower(), otherwise, the operators cannot be used. With these overload functions, fundamental algebraic computation of ppoly objects can be carried out directly.

It should be noted that, when writing the overload functions, they must be located in the folders beginning by @. They should not be placed anywhere else. Otherwise, the functions cannot be normally executed, and the other overload functions for other classes may also be affected.

For instance, if function plus() is located in the @ppoly folder, and we want to compute $p_1+p_2$, the MATLAB mechanism will find in the @ppoly folder the function plus(). If the function is found, it is executed, otherwise, an error message is given, indicating that addition is not defined. If the function is not placed in the @pploy folder, it cannot be found. More importantly, the execution of other classes may be affected, even the addition operation in MATLAB itself is overwritten, and unexpected errors may occur.

**Example 9.6.** Write an addition overload function for ppoly objects.

**Solutions.** For two given pseudo-polynomials, the addition operation is simple. The coefficient vectors of the two pseudo-polynomials can be cascaded together, we can also join the two order vectors, then, collecting the like terms, the simplified result is the sum of two pseudo-polynomials. Based on the idea, the overload function below can be written.

```
function p=plus(p1,p2)
p1=ppoly(p1); p2=ppoly(p2); a=[p1.a,p2.a]; na=[p1.na,p2.na];
p=ppoly(a,na); p=simplify(p);
```

where overload function simplify() should also be written, with the aim of collecting like terms in the ppoly objects, such that the simplest ppoly object can be obtained. The function will be written next.

### 9.3.2 Simplification functions via like-term collection

As it was pointed out earlier, if two pseudo-polynomials are added, the coefficient and order vectors should be joined together. It is inevitable that many like terms appears in the final pseudo-polynomials. The so-called like terms mean that two or more terms in the same pseudo-polynomial have the same order. A simple way to process them is to use like term collection methods. In the simplification function, the following points should be considered:

(1) Sorting the order vector in descending order to create a new vector, then taking the difference between adjacent elements. A loop structure can be used and if the difference term is zero (the difference is smaller than $10^{-10}$ in the program), such a term is a like term to the previous one, and the coefficient can be added to the previous one, and then the term is removed.

(2) After the loop structure, all the coefficients are examined. If the coefficients are zero (the absolute value is smaller than eps), such terms can be removed.

**Example 9.7.** Write a like-term collection function simplify() for the ppoly class.

**Solutions.** Based on the above considerations, a simplification function for pseudo-polynomials can be written as follows:

```
function p=simplify(p)
a=p.a; na=p.na;
[na,ii]=sort(na,'descend'); a=a(ii); ax=diff(na); key=1;
for i=1:length(ax)
    if abs(ax(i))<=1e-10,
        a(key)=a(key)+a(key+1); a(key+1)=[]; na(key+1)=[];
    else, key=key+1; end
end
ii=find(abs(a)>eps); a=a(ii); na=na(ii); p=ppoly(a,na);
```

**Example 9.8.** If $p_1(s) = 3s^{0.7} + 4s + 5$, $p_2(s) = 2s^{0.4} - 4s + 6s^{0.3} + 4$, compute $p_1(s) + p_2(s)$.

**Solutions.** The two pseudo-polynomials can be entered first. With the two overload functions, the sum of the two pseudo-polynomials can easily be computed. The final result obtained is $p(s) = 3s^{0.7} + 2s^{0.4} + 6s^{0.3} + 9$.

```
>> p1=ppoly([3 4 5],[0.7 1 0]);
   p2=ppoly([2 -4 6 4],[0.4 1 0.3 0]); p=p1+p2
```

### 9.3.3 Overload subtraction functions

For the minus sign, the corresponding function is `minus()`. An overload function should be written to define the subtraction action of ppoly objects. An example is given below to demonstrate the function programming.

**Example 9.9.** Write a minus overload function for the ppoly objects.

**Solutions.** When the addition method is defined, it is natural to represent the minus method $p(s) = p_1(s) - p_2(s)$ as the addition of $p(s) = p_1(s) + [-p_2(s)]$. However, the function $-p_2(s)$ needs its own overload functions. The function is referred to as unary minus function for ppoly object, with a fixed name `uminus()`. The signs of the coefficients are all altered, and the order is unchanged. The overload function can be written as

```
function p1=uminus(p)
p1=ppoly(-p.a,p.na); % the signs of coefficients are altered
```

With the `uminus()` overload function, the following function is written to define subtraction operations:

```
function p=minus(p1,p2)
p=p1+(-p2);   % substraction function based on addition
```

**Example 9.10.** For two pseudo-polynomials $p_1(s) = 3s^{0.7} + 4s + 5$ and $p_2(s) = 2s^{0.4} - 4s + 6s^{0.3} + 4$, compute $p_1(s) - p_2(s)$.

**Solutions.** With the overload functions, the following commands can be used to compute the subtraction of the two pseudo-polynomials, and the result is $p(s) = 8s + 3s^{0.7} - 2s^{0.4} - 6s^{0.3} + 1$.

```
>> p1=ppoly([3 4 5],[0.7 1 0]);
   p2=ppoly([2 -4 6 4],[0.4 1 0.3 0]); p=p1-p2
```

### 9.3.4 Overload multiplication functions

The multiplication sign $*$ is defined by the overload function `mtimes()`. In this section, an example is used to demonstrate the overload function programming. The definitions of Kronecker operations are given first.

**Definition 9.6.** For two matrices $A$ and $B$, the Kronecker product $A \otimes B$ is defined as

$$C = A \otimes B = \begin{bmatrix} a_{11}B & \cdots & a_{1m}B \\ \vdots & \ddots & \vdots \\ a_{n1}B & \cdots & a_{nm}B \end{bmatrix}. \tag{9.3.1}$$

**Definition 9.7.** The Kronecker sum $A \oplus B$ of two matrices $A$ and $B$ is defined as

$$D = A \oplus B = \begin{bmatrix} a_{11} + B & \cdots & a_{1m} + B \\ \vdots & \ddots & \vdots \\ a_{n1} + B & \cdots & a_{nm} + B \end{bmatrix}. \tag{9.3.2}$$

The MATLAB function $C$=`kron`$(A,B)$ can be used to compute the Kronecker product $A \otimes B$. A function `kronsum()` can be written to compute the Kronecker sum:

```
function C=kronsum(A,B)
[ma,na]=size(A); [mb,nb]=size(B);
A=reshape(A,[1 ma 1 na]); B=reshape(B,[mb 1 nb 1]);
C=reshape(bsxfun(@plus,A,B),[ma*mb na*nb]);
```

**Example 9.11.** Write an overload function for the product of two ppoly objects.

**Solutions.** Consider first the multiplication algorithm. If $p_1(s)$ is a ppoly object, each of its coefficients should be multiplied to all the coefficients in object $p_2(s)$. Therefore, Kronecker product should be used for the coefficient operation. For the order operation, each order in $p_1(s)$ is added to all the orders of $p_2(s)$. Therefore, Kronecker sum is suitable for the operation. Like term collection can be performed on the final result. Based on the considerations, the following overload function can be written:

```
function p=mtimes(p1,p2)
p1=ppoly(p1); p2=ppoly(p2); a=kron(p1.a,p2.a);
na=kronsum(p1.na,p2.na); p=simplify(ppoly(a,na));
```

Note that, in the first sentence, ppoly() function is used to make sure that the two input arguments are expressed as ppoly objects. With the ppoly.m file given earlier, the conversion to ppoly object is allowed by the function call.

**Example 9.12.** For the pseudo-polynomials $p_1(s) = 3s^{0.7} + 4s + 5$ and $p_2(s) = 2s^{0.4} + 6s + 6s^{0.3} + 4$, compute $p_1(s)p_2(s)$.

**Solutions.** When the multiplication overload function is written, the following statements can be used to enter the two ppoly objects, and compute the product of them:

```
>> p1=ppoly([3 4 5],[0.7 1 0]);
   p2=ppoly([2 6 6 4],[0.4 1 0.3 0]); p=p1*p2
```

The result obtained is

$$p(s) = 24s^2 + 18s^{1.7} + 8s^{1.4} + 24s^{1.3} + 6s^{1.1} + 64s + 12s^{0.7} + 10s^{0.4} + 30s^{0.3} + 20.$$

**Example 9.13.** For the given pseudo-polynomials $p_1(s)$ and $p_2(s)$ in Example 9.12, compute $p(s) = p_1^4(s)p_2^2(s)$.

**Solutions.** With the defined mtimes() overload function, the following continuous multiplication can be carried out:

```
>> p1=ppoly([3 4 5],[0.7 1 0]);
   p2=ppoly([2 6 6 4],[0.4 1 0.3 0]);
   p=p1*p1*p1*p1*p2*p2
```

The result obtained is

$$
\begin{aligned}
p(s) = {} & 9\,216s^6 + 27\,648s^{5.7} + 37\,248s^{5.4} + 18\,432s^{5.3} + 33\,984s^{5.1} + 113\,664s^5 + 24\,676s^{4.8} \\
& + 208\,896s^{4.7} + 9\,216s^{4.6} + 13\,440s^{4.5} + 203\,584s^{4.4} + 132\,096s^{4.3} + 5\,400s^{4.2} \\
& + 148\,152s^{4.1} + 427\,264s^4 + 1\,728s^{3.9} + 85\,040s^{3.8} + 523\,392s^{3.7} + 46\,404s^{3.6} \\
& + 33\,336s^{3.5} + 375\,836s^{3.4} + 337\,920s^{3.3} + 9\,936s^{3.2} + 221\,040s^{3.1} + 682\,880s^3 \\
& + 2\,160s^{2.9} + 90\,816s^{2.8} + 550\,800s^{2.7} + 86\,400s^{2.6} + 23\,040s^{2.5} + 298\,080s^{2.4} \\
& + 388\,800s^{2.3} + 5\,400s^{2.2} + 134\,640s^{2.1} + 486\,300s^2 + 29\,600s^{1.8} + 242\,400s^{1.7} \\
& + 72\,000s^{1.6} + 6\,000s^{1.5} + 104\,600s^{1.4} + 195\,000s^{1.3} + 24\,000s^{1.1} + 134\,000s \\
& + 2\,500s^{0.8} + 39\,000s^{0.7} + 22\,500s^{0.6} + 10\,000s^{0.4} + 30\,000s^{0.3} + 10\,000.
\end{aligned}
$$

### 9.3.5 Overload power functions

MATLAB overload function mpower() can be written to compute the power of a pseudo-polynomial. With such a function, the operator ^ can be used in MATLAB expression. Again the class ppoly is used to demonstrate the power overload functions.

**Example 9.14.** For the given class ppoly, write the power overload function.

**Solutions.** Consider the descriptions in Theorem 9.2. The following two points are considered in the programming of the overload functions:
(1)  If $p$ is an $s$ operator, $n$ can be any real number, with the result $s^n$;
(2)  If $p$ is a ppoly object, $n$ must be an integer, otherwise, an error message is given.

In fact, (1) can further be extended as: if $p$ is a pseudo-polynomial object, and it has only one term, the power can be taken for any real number $n$. With the factors considered, the overload function can be written as

```
function p1=mpower(p,n)
if length(p.a)==1, p1=ppoly(p.a^n,p.na*n);
elseif n==floor(n)
    if n<0, p.na=-p.na; n=-n; end
    p1=ppoly(1); for i=1:n, p1=p1*p; end
else, error('n must be an integer'), end
```

**Example 9.15.** Input the pseudo-polynomial $p(s) = 3s^{0.7} + 4s + 5$ in an alternative way in Example 9.5.

**Solutions.** If the pseudo-polynomial in Example 9.5 is to be entered, the two vectors should be extracted first. An alternative way is considered here with the overload operations defined above. The operator $s$ can be defined, and through simple expression of the pseudo-polynomials, the following statements can be used:

```
>> s=ppoly('s'); p=3*s^0.7+4*s+5
```

With the above expression, MATLAB mechanism can be used to compute the desired operations for the ppoly object. The final result is a ppoly object, which is the same as in Example 9.5.

**Example 9.16.** Solve the problem in Example 9.13 with power functions.

**Solutions.** Two methods can be used to compute the expression $p_1^4(s)p_2^2(s)$, the results of continuous multiplication of the power function are exactly the same, and the difference of the two expressions is zero.

```
>> p1=ppoly([3 4 5],[0.7 1 0]);
   p2=ppoly([2 6 6 4],[0.4 1 0.3 0]);
   p0=p1*p1*p1*p1*p2*p2, p=p1^4*p2^2, p-p0
```

**Example 9.17.** For the given pseudo-polynomial $p(s) = s^{0.7} - 2s + 3$, compute $p^{-3}(s)$.

**Solutions.** The orders defined here are not restricted to nonnegative values. Therefore, it is also possible to compute $p^{-3}(s)$. The result can be obtained with the following commands:

```
>> p=ppoly([1 -2 3],[0.7 1 0]); p1=p^-3
```

The result obtained is

$$p_1(s) = 27 + 27s^{-0.7} - 54s^{-1} + 9s^{-1.4} - 36s^{-1.7} + 36s^{-2} + s^{-2.1} - 6s^{-2.4} + 12s^{-2.7} - 8s^{-3}.$$

### 9.3.6 Assignment and extraction of fields

If the name of the object is $p$, two methods can be used to extract the fields. The first method is to use $p.a$ and $p.na$ commands to extract the fields. The other method is to use get$(p,$'a'$)$ and get$(p,$'na'$)$ commands. In the former syntax, the first method is automatically supported. While the latter needs programming. The following overload function can be written.

```
function p1=get(varargin)
p=varargin{1};
if nargin==1,
    s=sprintf('%f ,',p.a); disp(['   a: [' s(1:end-1) ']'])
    s=sprintf('%f ,',p.na); disp([' na: [' s(1:end-1) ']'])
elseif nargin==2, key=varargin{2};
    switch key,
        case 'a', p1=p.a; case 'na', p1=p.na;
        otherwise, error('Wrong field name used'), end
else, error('Wrong number of input arguments'); end
```

Similarly, another overload function set() should be written. In the object ppoly, only two fields are involved. It is not necessary to write such a function. The user may try to write the function in the problems.

## 9.4 Inheritance and extension of classes

The inheritance of classes is also an important topic in object-oriented programming. The related definition is proposed as follows.

**Definition 9.8.** If class A is inherited from another class B, then A is referred to as a subclass of B, and B is referred to as a parent of class A.

Inheritance may enable the subclass have all the properties and methods of the parent class. There is no need to write repetitive code. In this section, the extension technique of classes is mainly demonstrated.

### 9.4.1 Definition and display of extended classes

In practical object-oriented programming, a higher-level class can be defined on top of an existing class. In this section, the extension of a class is illustrated through examples.

**Definition 9.9.** The mathematical form for a fractional-order transfer function is defined as

$$G(s) = \frac{b_1 s^{\beta_1} + b_2 s^{\beta_2} + \cdots + b_m s^{\beta_n}}{a_1 s^{\alpha_1} + a_2 s^{\alpha_2} + \cdots + a_n s^{\alpha_n}} \tag{9.4.1}$$

where the numerator and denominator are both pseudo-polynomials, denoted respectively by $N(s)$ and $D(s)$.

A complete design of fractional-order transfer function model is given in [19], with the FOTF Toolbox. The design process is not going to be repeated here. A simple method can be introduced to redesign some of the overload functions presented here. For better use of fractional-order system modeling, analysis and design tasks, the users are strongly recommended to use FOTF Toolbox instead[20].

**Example 9.18.** Based on (9.4.1), design a new MATLAB class with extension of ppoly objects.

**Solutions.** In the FOTF Toolbox, the class name is selected as fotf. In order to show the difference between the well-established FOTF Toolbox, a new class can be named as ftf. A folder @ftf should be created first, where the fields $N(s)$ and $D(s)$ are, in fact, two pseudo-polynomials. These two variables can be designed as the fields of the object, with names pnum and pden, respectively, of ppoly objects.

```
classdef ftf
    properties, pnum, pden, end
    methods
```

```
function G=ftf(a,na,b,nb)
switch nargin
   case 1,
      if isa(a,'double'), G=ftf(ppoly(a),ppoly(1));
      elseif isa(a,'ppoly'), G=ftf(a,ppoly(1));
      elseif a=='s', G=ftf(ppoly('s'),ppoly(1)); end
   case 2,
      if isa(a,'ppoly'), G.pnum=a; G.pden=ppoly(na); end
   case 3,
      if isa(a,'ppoly'), b=ppoly(na,b); G=ftf(a,b);
      else, G=ftf(ppoly(a,na),ppoly(b)); end
   case 4, G=ftf(ppoly(a,na),ppoly(b,nb));
end, end, end, end
```

In the display.m file of ppoly class, a returned argument was reserved. Now it can be used in designing the display function for the ftf class. The display of an ftf object is, in fact, the display of two ppoly objects, and in-between a horizontal line is displayed. Therefore, the following function is designed:

```
function display(G)
strN=display(G.pnum); strD=display(G.pden);
nn=length(strN); nd=length(strD); nm=max([nn,nd]);
disp([char(' '*ones(1,floor((nm-nn)/2))) strN])
disp([char('-'*ones(1,nm))]);
disp([char(' '*ones(1,floor((nm-nd)/2))) strD])
```

It can be seen that with the support of the low-level ppoly object, the overload function programming of the new object can be established upon the definition of the ppoly object. Therefore, it is more concise than that in the FOTF Toolbox.

**Example 9.19.** Input the following fractional-order transfer function into MATLAB:

$$G(s) = \frac{-2s^{0.63} - 4}{2s^{3.501} + 3.8s^{2.42} + 2.6s^{1.798} + 2.5s^{1.31} + 1.5}.$$

**Solutions.** The coefficient and order vectors of the denominator and numerator can be retrieved from the given model, then an ftf object can be established with

```
>> a=[-2 -4]; na=[0.6 0]; b=[2 3.8 2.6 2.5 1.5];
   nb=[3.501 2.42 1.798 1.31 0]; G=ftf(a,na,b,nb)
```

The model entered can be displayed directly with

$$-2*s^{0.6}-4$$
$$\overline{\phantom{-----------------------------------------------------------}}$$
$$2*s^{3.501}+3.8*s^{2.42}+2.6*s^{1.798}+2.5*s^{1.31}+1.5$$

The model can alternatively be entered with the following statements, and the same model can be obtained:

```
>> num=ppoly([-2 -4],[0.6 0]);
   den=ppoly([2 3.8 2.6 2.5 1.5],[3.501 2.42 1.798 1.31 0]);
   G1=ftf(num,den)
```

### 9.4.2 Overload functions for ftf objects

It can be seen from the definition of the ftf object that, when addition, subtraction, multiplication, and division are computed, the result is still an ftf object. The algebraic operations of lower-level ppoly objects can be performed to complete the result ftf object. The ftf object is still used as an example to demonstrate the algebraic operations of the fractional-order transfer function class.

**Definition 9.10.** If two ftf objects, denoted respectively as $G_1(s) = N_1(s)/D_1(s)$ and $G_2(s) = N_2(s)/D_2(s)$, are added, the resulted ftf object can be obtained from the operations of the ppoly objects as

$$G_1(s) + G_2(s) = \frac{N_1(s)}{D_1(s)} + \frac{N_2(s)}{D_2(s)} = \frac{N_1(s)D_2(s) + N_2(s)D_1(s)}{D_1(s)D_2(s)} \tag{9.4.2}$$

where $N_i(s)$ and $D_i(s)$ are all ppoly objects.

**Example 9.20.** Write the overload addition function for the ftf class.

**Solutions.** With the overload function of the addition defined below, it can be seen that the function itself is concise and straightforward. The sum and product functions defined in ppoly are used directly. The extended programming is simple and convenient, and no low-level programming is needed.

```
function G=plus(G1,G2)
G1=ftf(G1); G2=ftf(G2); N1=G1.pnum; D1=G1.pden;
N2=G2.pnum; D2=G2.pden; G=ftf(N1*D2+N2*D1, D1*D2);
```

**Definition 9.11.** Similarly, the multiplication formula for ftf objects can be expressed as

$$G_1(s)G_2(s) = \frac{N_1(s)N_2(s)}{D_1(s)D_2(s)}. \tag{9.4.3}$$

**Example 9.21.** Write a multiplication overload function for the ftf class.

**Solutions.** Based on the above formula, the overload function `mtimes()` can be written to compute directly the product of two ftf objects. Objects ppoly are used again to avoid low-level computation.

```
function G=mtimes(G1,G2)
G1=ftf(G1); G2=ftf(G2); N1=G1.pnum; D1=G1.pden;
N2=G2.pnum; D2=G2.pden; G=ftf(N1*N2, D1*D2);
```

**Definition 9.12.** If the ftf of the forward path is given by $G_1(s)$, while that in the feedback path is $G_2(s)$, the overall model of the negative feedback system can be computed from

$$G(s) = \frac{G_1(s)}{1 + G_1(s)G_2(s)} = \frac{N_1(s)D_2(s)}{N_1(s)N_2(s) + D_1(s)D_2(s)}. \tag{9.4.4}$$

**Example 9.22.** Write an overload function for the feedback connection of ftf objects.

**Solutions.** Based on the above mentioned mathematical formula, the overload function `feedback()` of two ftf objects can be written as

```
function G=feedback(G1,G2)
G1=ftf(G1); G2=ftf(G2); N1=G1.pnum; D1=G1.pden;
N2=G2.pnum; D2=G2.pden; G=ftf(N1*D2, N1*N2+D1*D2);
```

In the previous three functions, `mtimes()` can be used to find the overall model in series connection, `plus()` computes the overall model for parallel connection, and `feedback()` computes the feedback connection models. With these three functions, fractional-order transfer function models under complicated connections can be computed easily.

**Example 9.23.** Write the other overload functions for algebraic operations of ftf objects.

**Solutions.** Similar to the unary minus function and minus function of the ppoly objects, the overload functions can be written respectively as

```
function G1=uminus(G)
G1=ftf(-G.pnum,G.pden);
```

```
function G=minus(G1,G2)
G=ftf(G1)+(-ftf(G2));
```

Similar to the `mpower()` function of ppoly object, an overload function `mpower()` can be designed for ftf objects as

```
function G1=mpower(G,n)
p1=G.pnum; p2=G.pden;
if length(p1.a)==1 & length(p2.a)==1, G1=ftf(p1^n,p2^n);
elseif n==floor(n) & n>=0,
   G1=ftf(1); for i=1:n, G1=G1*G; end
else, error('n must be an integer'), end
```

The division overload functions can also be defined for the ftf objects. For the division sign /, a right division function mrdivide() can be written. The division operation is not supported in ppoly class, since the result is no longer a ppoly object. However, for ftf objects, since the division result is still an ftf object, the division overload function can be written as

```
function G=mrdivide(G1,G2)
G1=ftf(G1); G2=ftf(G2); N1=G1.pnum; D1=G1.pden;
N2=G2.pnum; D2=G2.pden; G=ftf(N1*D2, N2*D1);
```

**Example 9.24.** Input the fractional-order transfer function model in Example 9.19 in an alternative way.

**Solutions.** With s defined as an ftf operator, with the above mentioned overload functions, the fractional-order transfer function model can be entered into MATLAB environment. It can be seen that the result is exactly the same as that in Example 9.19.

```
>> s=ftf('s');
   G2=(-2*s^0.6-4)/(2*s^3.501+3.8*s^2.42+2.6*s^1.798+2.5*s^1.31+1.5)
```

### 9.4.3 Frequency domain analysis of fractional-order transfer functions

For a given model $G(s)$, the operator $s$ is substituted by a frequency vector $j\omega$, so a complex vector $G$ can be obtained. From the complex vector and the frequency vector, Bode diagram of the function can be immediately computed. The overload function bode() for Bode diagram is demonstrated through examples.

**Example 9.25.** Write an overload function for Bode diagram of a fractional-order transfer function models. Draw the Bode diagram of the system in Example 9.19.

**Solutions.** If for a pseudo-polynomial object, a computing function can be written for frequency domain response computation, then an overload function for the Bode diagram can be written for fractional-order transfer function models.

Considering first the frequency response computation function for a ppoly object $p$, the function should be placed in the @ppoly folder.

```
function H=freqw(p,w)
for i=1:length(w), H(i)=p.a(:).'*[1i*w(i)].^p.na(:); end
```

Based on the low-level function `freqw()`, an overload function for Bode diagrams can be computed and obtained as shown below.

```
function H=bode(G,w)
H=freqw(G.pnum,w)./freqw(G.pden,w);
subplot(211), semilogx(w,20*log10(abs(H)))
subplot(212), semilogx(w,angle(H)*180/pi)
```

For the fractional-order transfer function model studied in Example 9.19, the model $G(s)$ can be entered first, and then the Bode diagram can be drawn with overload function `bode()`, as shown in Figure 9.1.

```
>> a=[-2 -4]; na=[0.6 0]; b=[2 3.8 2.6 2.5 1.5];
   nb=[3.5 2.4 1.8 1.3 0]; G=ftf(a,na,b,nb);
   w=logspace(-2,2,100); bode(G,w);
```

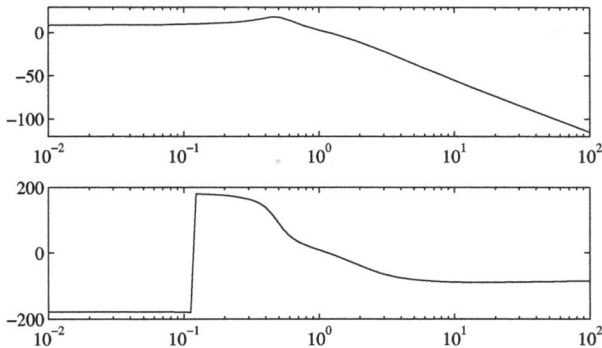

**Figure 9.1:** Bode diagram for a fractional-order transfer function model.

## 9.5 Problems

9.1 For a newly defined operation, $a \times b = (a + b) + 2(a - b)$, where the operator $\times$ is redefined, design a new object, and compute under the new operator the expressions $123 \times 54$ and $123 \times (32 \times 23)$.

In fact, for this particular operation, even though the class is not defined, a simpler solution to the problem can be made, by writing out a new function `new-times()`, or an anonymous function to define the new operation. The method is much simpler, however, the new operation expressed cannot be obtained with the operator *.

9.2 Write an overload function `latex()` for the ppoly object, such that the display of the object can be expressed directly into a LaTeX string.

9.3 For a ppoly object, write an overload function for the operator ==, whose MATLAB function name is `eq()`, to test whether two ppoly objects are the same. If they are, a logic 1 is returned, otherwise, logic 0 can be obtained.

9.4 For ppoly objects, write an overload function `set()`, with the syntax
`set(p, property name, property value)`
It can also used to define several property names together. If the field name is not `'a'` or `'na'`, an error message should be displayed.

9.5 Using inheritance, define a subclass for the existing ppoly class such that only nonnegative order vector $\alpha$ is allowed.

9.6 For an ftf object, write an overload function `latex()` such that the ftf object can be converted into a LaTeX string.

9.7 Write overload functions for an ftf object such that Nyquist plots and Nichols charts can be drawn.

9.8 For an ftf object, write an overload function `inv()` such that the reciprocal of the fractional-order transfer function model can be obtained. Hints: the inverse of fractional-order transfer function model is, in fact, the swap of the numerator and denominator information.

9.9 Modify the overload function `bode()` in an ftf object such that no sudden change appears in the phase plot.

# 10 Graphical user interface design using MATLAB

For a successful software, of course, its contents and functions are the most important factors. The graphical user interface is also very important since it determines the quality and level of the software. The graphical user interface (GUI) in this case represents the outward appearance of the product. Thus, by mastering the skills of graphical user interface techniques, it is possible to design high-quality software for general purposes.

Some ideas about handle graphics were introduced earlier. The properties of some objects were accessed with appropriate statements. In practical graphical interface design, it is better to use the handy tool called Guide. However, Guide is not everything. Sometimes low-level statements are necessary in graphical user interface design. In this chapter, low-level statements are presented first, then Guide program is illustrated such that the design of interfaces may be improved.

In Section 10.1, fundamental information of graphical interface programming is presented, and the relationships among various objects are introduced. The setting and extraction of window object properties are discussed, modification methods of the properties of objects are illustrated. Some standard dialog boxes in MATLAB are also elaborated. Commonly used controls are discussed in Section 10.2, and also the methods of how to extract the handles are presented. In Section 10.3, a visual design tool Guide is presented, and with the tool, the design of GUIs is demonstrated with examples. In Section 10.4, advanced techniques in graphical user interface design are illustrated. The design of menu systems and toolbars are also presented. Examples are given to illustrate the design of ActiveX controls in interfaces. In Section 10.5, the integration and packaging of user-designed toolboxes are presented.

## 10.1 Essentials in graphical user interface design

### 10.1.1 The relationships of objects in MATLAB interface

Various objects are supported in MATLAB graphical user interfaces. The relationships of the objects are discussed in Figure 10.1. There is a root object in MATLAB environment, and its handle value is 0. Based on the root object, the figure window objects are defined. The values of the figure window handles are positive integers.

For each graphical window object, there are four lower-level objects, such as the menu, control, axis, and shortcut menu objects. Menu objects can be used to design the main menu system of the interface. Control objects are responsible for representing various controls in the interfaces. The controls include buttons, edit boxes, listboxes, and so on. The shortcut menu objects allow the user to create a shortcut menu for the interface, so that right clicking the mouse may bring in the shortcut menus. Theoretically, each control may have its own shortcut menu, while in real applications, since

https://doi.org/10.1515/9783110666953-010

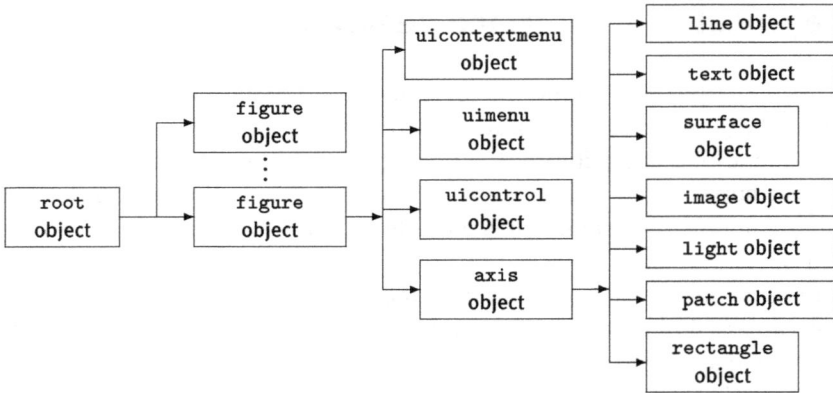

**Figure 10.1:** Illustration of the existing figure objects.

the support in MATLAB is not fully provided, some of the shortcut menus cannot be activated.

For the axis object, various lower-level objects are allowed. Some information about them was presented earlier. In this chapter, a detailed account is presented on property setting, menu system and control design in graphical user interfaces, so that the readers have better understanding on the concept and applications of handle graphics.

### 10.1.2 Window objects and properties

If one wants to open a new window in MATLAB, the menu item New → Figure Window in the command window can be clicked, so that a standard MATLAB graphics window is opened. Besides, the following command can be used to open a graphics window:

hwin=figure(property 1, value 1, property 2, value 2, ... )

where hwin is the handle of the window. It is obvious that the window can be manipulated through the handles, so as to extract and modify the properties of the window. When a window is opened, the window handle hwin can be obtained. Command figure(hwin) can be used to bring the window to the front. In fact, even though the handle does not exist, a new figure window can be opened, and defined as the current window. The value of the figure window handle is a positive integer.

### 10.1.3 Commonly used properties in window objects

Many windows are allowed to be opened by the user in MATLAB environment, with each window having its own handle. Function gcf() can be used to get the handle of the current window, so as to operate the window. If the menu item View → Property Inspector is selected, a dialog box shown in Figure 10.2 can be opened. If a certain property in the dialog box is selected, the property value can be filled in. The commonly used properties are listed below:

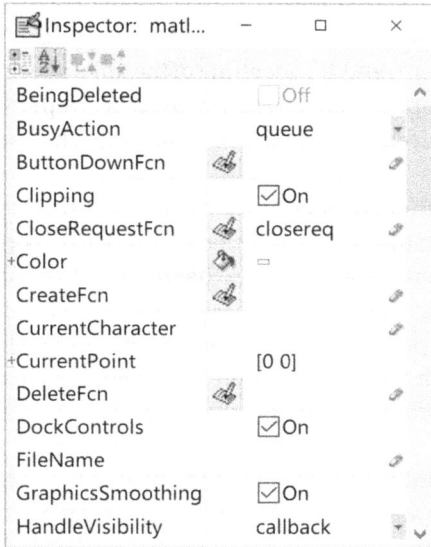

**Figure 10.2:** Property editor.

- Color property is used to set the background color of the window. The property value is a $1 \times 3$ row vector storing the components of the three primitive colors, with the range of 0 to 1. The combinations of the eight commonly used colors are show in Table 6.1. For instance, the constant 'r' corresponds to the vector $[1, 0, 0]$, the red color. When the color resolution of the Windows environment is high, arbitrary combination between 0 and 1 can be used, and the interface is colorful. If one wants to set the background color to $[0.8, 0.8, 0.8]$, one can click the button labeled "..." in the Color property. A standard color dialog box can be displayed so that the appropriate color can be chosen.
- MenuBar property is used in setting the format of the menu system. The available options are 'figure' or 'none', with the former being the default one. If 'none' option is selected, the initial menu system is defined, and the menu system of the window can be set with function uimenu(), to be illustrated later in the chapter. If the option 'figure' is selected, the default figure window menu system is inherited. Other menu items can be appended further with the function uimenu().
- Name property sets the title bar of the figure window. The property value is a string, and the information will be set to the window title bar.

- NumberTitle property is used to determine whether the figure number appears in the title of the window. The available options are 'on' and 'off', with the former being the default one. If the 'on' option is selected, the words **Figure No \*:** appear in the title bar of the window. If the option 'off' is used, the figure number will not appear in the window title.
- Units property allows the user to set the unit of the length, with the default option of 'pixels'. Besides, the options 'inches', 'centimeters' and 'normalized' can also be used, with the latter set between 0 and 1. The setting of the units affects the properties such as **Position**. Units property can also be set with the property inspector interface. If **Units** property is selected, a listbox can be opened, and the desired property can be selected.
- Position property is used to set the position and size of the figure window. The value of the property is a $1 \times 4$ vector, with the first two elements setting the coordinates of the lower left corner, and the remaining two values the width and height of the window, respectively. The units of the lengths are associated with the **Units** property. The best way to set **Position** property is by first closing the property setting dialog box, resizing the window with the mouse button, and then, opening the property inspector program. The appropriate sizes can be set in the **Position** edit box.
- Resize property is used to determine whether the size of the window can be adjusted or not. There are two options, 'on' and 'off', which can be selected, with 'on' being the default one.
- Toolbar property indicates whether toolbar is added to a window or not. The options are 'none', 'figure' (with standard toolbox), and 'auto'. The default option is 'figure'.
- Visible property is used to determine whether the window established is visible or not. The corresponding values are 'on' and 'off', with 'on' being the default value.
- Pointer property can be used to set the shape of the cursor, and the values of the properties shown in Table 10.1, with 'arrow' being the default option.
- Various callback functions. The so-called callback function means that, if an event happens, the function is invoked. The functions can be expressed in strings, or

**Table 10.1:** Property values for Pointer shapes.

| name | cursor shape | name | cursor shape |
| --- | --- | --- | --- |
| 'crosshair' | thin cross sign | 'arrow' | arrow pointer |
| 'watch' | meaning waiting | 'circle' | round shaped indicator |
| 'topl' | right upward arrow | 'botl' | similar to 'topr', left upward arrow |
| 'topr' | right downward arrow | 'botr' | similar to 'topl', right upward arrow |
| 'cross' | double cross sign | 'fleur' | cross sign with arrows |

given in MATLAB files. If can also be a group of statements in MATLAB. Note that the callback functions should be a string which can be evaluated. In graphics windows, the commonly used callback functions are:

- CloseRequestFcn, the response function when the window is closed;
- KeyPressFcn, the function when a key is pressed;
- WindowButtonDownFcn, the function when a mouse button is pressed;
- WindowButtonMotionFcn, the function when a mouse is moved;
- WindowButtonUpFcn, the function when a mouse button is released;
- CreateFcn and DeleteFcn, the functions when an object is created or deleted;
- CallBack, the function when an object is selected.

The property values can be set to MATLAB function names or commands, meaning that if an event happens, the corresponding callback function is triggered.

### 10.1.4 Extraction and modification of object properties

The properties and their values can be set up when the window object is created. Also they can be assigned in the program dynamically, with the functions set() and get(), introduced earlier. If one wants to modify certain properties of an object, set() function can be called in MATLAB, with the syntax

set(handle,name 1,value 1,name 2,value 2,...)

where the handle of the object to be modified must be provided. Note that when function set() is used, the property names should be quoted with single quotation marks. If the property values are strings, they should also be quoted with single quotation marks. In the set() function call, with no property values assigned, all the available property values are listed. For instance, if the following statement is used, the result is returned in the cell array argument h

```
>> h=set(gcf,'Visible')
```

where the returned h is a cell array, with {'on','off'}. The get() function is called with the following syntax:

V=get(handle,property)

where V is the returned property value. Examples are used to demonstrate the modification of window properties.

**Example 10.1.** Open a blank window. Display "Hello, keyboard key pressed" when any key is pressed.

**Solutions.** A blank window can be opened with the following statements:

```
>> gwin=figure('Visible','off');
   set(gwin,'Color',[1,0,0],'Position',[100,200,300,300],...
       'Name','My Program','NumberTitle','off','MenuBar','none',...
       'KeyPressFcn','disp(''Hello, keyboard key pressed'')');
   set(gwin,'Visible','on')
```

The window was opened first as an invisible one, and returned a handle gwin. Then, set() could be used to assign the background color of the window to red, and the initial position was also assigned, with the default unit being pixel. A title of the window was set to "My Program", and the initial menu system was canceled.

A callback function is assigned to the window, in response to the **KeyPressFcn** property. If a key is pressed, the MATLAB mechanism invokes the callback function assigned to it. The function disp() embedded in the callback function is executed, such that the line "Hello, keyboard key pressed" is displayed in the MATLAB command window.

Pay attention to the description in the callback function. The actual callback function is disp('Hello, keyboard key pressed'), however, since the whole command should be quoted again by single quotation marks, the original quotation marks inside the string should be replaced by two consecutive single question marks.

After a blank window is prepared, function set() is used again to make it visible.

If the following statements are written, the position and color of the windows are returned in vectors $v_1 = [100, 159, 300, 400]$ and $v_2 = [1, 0, 0]$.

```
>> v1=get(gwin,'Position'), v2=get(gwin,'Color')
```

**Example 10.2.** Modify the properties of the interface with set() function.

**Solutions.** It can be seen that function set() can be used to extract or modify the properties of Windows interface easily, and the result can be displayed immediately. The following seeming simpler code can be used:

```
>> gwin=figure;
   set(gwin,'Color',[1,0,0],'Position',[100,200,300,300],...
       'Name','My Program','NumberTitle','off','MenuBar','none',...
       'KeyPressFcn','disp(''Hello, Keyboard key pressed'')');
```

When the window is opened. It seems that the two codes yield the same results. The windows are opened and their handles are assigned to the variable gwin. However, the intermediate processes are different. In the latter command, the modification process is all displayed, while in the former, the modification processes are invisible, and the modified window appears as if a new window was opened. In real applications, the former approach is recommended. In fact, a simpler command can be used:

```
>> gwin=figure('Color',[1,0,0],'Position',[100,200,300,300],...
      'Name','My Program','NumberTitle','off','MenuBar','none',...
      'KeyPressFcn','disp(''Hello, keyboard key pressed'')');
```

to replace the above statements, and the effect is identical. The latter one is more concise and simpler.

### 10.1.5 Easy dialog boxes

Some easy dialog boxes are provided in MATLAB, and can be called in a simple manner. In this section, the use of some of the standard dialog boxes is demonstrated.

(1) **Message display box.** Function msgbox(string,title) can be used to display messages directly.

(2) **Warning or error message boxes.** Such messages can be handled with warndlg() and errordlg() functions. The differences between the two are the icons and color.

**Example 10.3.** Compare the behaviors of the warning and error boxes.

**Solutions.** The following MATLAB command can be used:

```
>> h=warndlg({'ERROR: There is ... encountered',...
      'Try again'},'Warning')
```

A warning message box is displayed in Figure 10.3(a). Multiple lines of information are supported in the box. Here, a cell array is used as a solution. Characters of other languages can be used as well.

If the following statements are used:

```
>> h=errordlg({'ERROR: There is ... encountered',...
      'Try again'},'Error')
```

the error message box is opened as shown in Figure 10.3(b).

(3) **Question and answer boxes.** MATLAB provides a function questdlg() which can be to give prompts, and allows the user to answer **Yes**, **No**, or **Cancel**. The syntax of the function is

```
key=questdlg(question string,title)
```

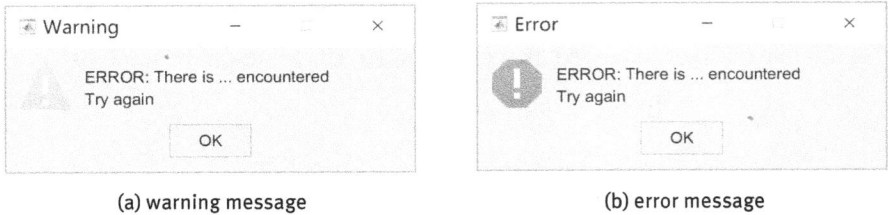

(a) warning message            (b) error message

**Figure 10.3:** Warning and error message boxes.

The return argument `key` can be a string `'Yes'`, `'No'`, or `'Cancel'`. Other strings are also allowed, as shown in the following example.

**Example 10.4.** Design a question dialog box with the label of Do you really want to quit?

**Solutions.** The following statements can be used to open such a dialog box, as shown in Figure 10.4(a). If a button is clicked, the dialog box is automatically closed, with the selected string, `'Yes'`, `'No'`, or `'Cancel'`, assigned to the returned argument `key`.

```
>> yesno=questdlg('Do you really want to quit?','Answer')
```

If one wants to change the text on the buttons, as `'yes'` and `'no'`, two buttons can be designed with labels **yes** and **no**, with the last `'no'` indicating that the button **no** is the default one, as shown in Figure 10.4(b). The selected result `'Yes'` or `'No'` is automatically assigned to the returned argument `key`.

```
>> yesno=questdlg('Really quit?','Answer','yes','no','no')
```

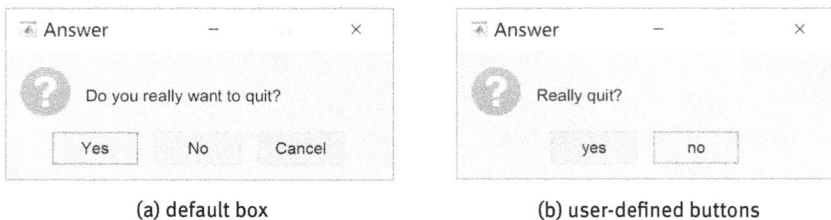

(a) default box            (b) user-defined buttons

**Figure 10.4:** Example of question-and-answer box.

(4) Variable input box, with a syntax of

variable=inputdlg({prompt 1,..., prompt $n$}, title, rows, default values)

where "prompt $i$" is the string for the $i$th variable to input; "title" is the string of the title of the dialog box; "rows" is the maximum number of rows for each variable; "default values" is a cell array, containing all the default values of the input variables. The returned "variable" is a cell string, containing the value of the $i$th variable in the $i$th cell of the array. The variables are in strings.

**Example 10.5.** Design a transfer function input dialog box for a control system.

**Solutions.** A transfer function of a control system is usually a rational expression, while its numerator and denominator are both polynomials. For polynomials, the coefficient vectors in descending order of $s$ can be used. The dialog box in Figure 10.5 can be designed with the following statements. If the transfer function is successfully entered, the vectors can be extracted from the strings, and with the function tf() in the Control System Toolbox, the transfer function object of a system can be constructed.

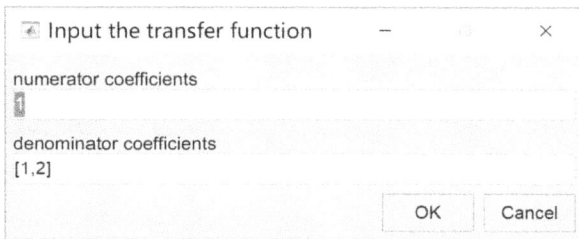

**Figure 10.5:** Dialog box of a transfer function model.

```
>> P={'numerator coefficients','denominator coefficients'};
   T=inputdlg(P,'Input the transfer function',1,{'1','[1,2]'});
   num=eval(T{1}); den=eval(T{2}); G=tf(num,den)
```

## 10.1.6 Standard dialog boxes

To make the graphical interface look like standard, some standard dialog boxes can be opened directly in MATLAB. The syntaxes of these functions are simple and straightforward. Some of the functions are given below:

(1) **File name dialog boxes.** Functions uigetfile() and uiputfile() can be used to open a standard dialog box, prompting the users to select a file name. The dialog boxes are used respectively for read and write files. Considering the former function, the syntax is

[file name, path name]=uigetfile(file name filter, title, $x$, $y$)

For instance, if one wants to open *.m files, the "file name filter" can be set to '*.m'. If more suffixes are allowed, they should be separated by semicolons. The "title" is a string variable with the title name of the dialog box. The values $x$ and $y$ are used to set the positions of the dialog box, and they can be omitted for default positions.

The returned "file name" and "path name" are the actual names expressed in strings. If no file name is specified, the returned "file name" is 0.

**Example 10.6.** Select a file from a standard dialog box such that the file and path names can be obtained.

**Solutions.** With the following MATLAB statement:

```
>> [f,p]=uigetfile('*.m;*.txt;*.c','Please select a file name')
```

a dialog box shown in Figure 10.6 can be opened. If a file is selected from the dialog box, then the file name $f$ and path name $p$ can be returned as strings. If Cancel button is pressed, the returned arguments are $f = p = 0$.

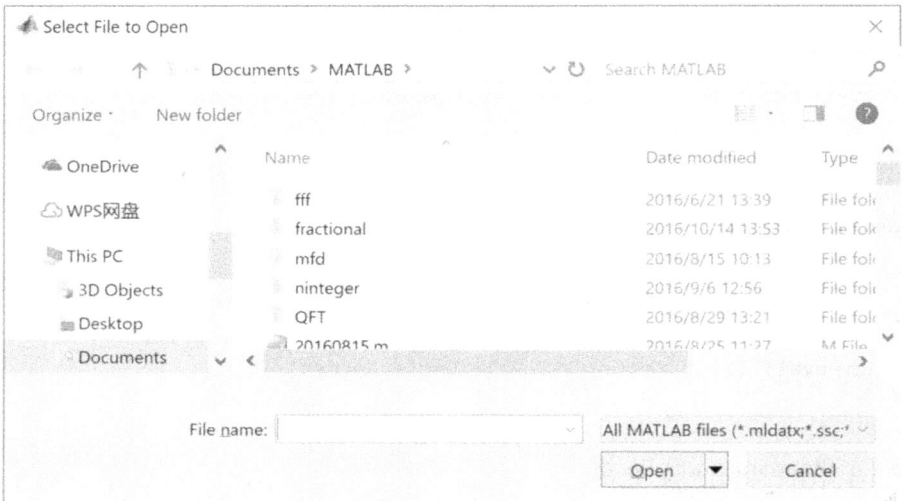

**Figure 10.6:** File name selection dialog box.

Besides, when `uiputfile()` function is called, and the file name specified exists, another dialog box is opened to ask whether the file is to be overwritten.

(2) Color dialog box. The standard color setting dialog box can be opened with function `uisetcolor()`, and the syntax is

$c$=uisetcolor; or $c$=uisetcolor($c_0$);

where in the first command, the dialog box opened is shown in Figure 10.7(a), and the returned argument is a $1 \times 3$ vector, storing the components of red, green, and blue primitive colors. The values of the components are within 0 and 1. If a vector

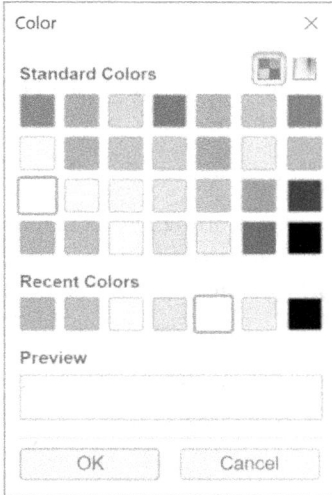

(a) color setting          (b) font setting

**Figure 10.7:** Color and font setting dialog boxes.

$c_0$ is supplied, the color in the dialog box points to that specified by the initial color vector $c_0$. If Cancel button is pressed, an empty vector is returned.

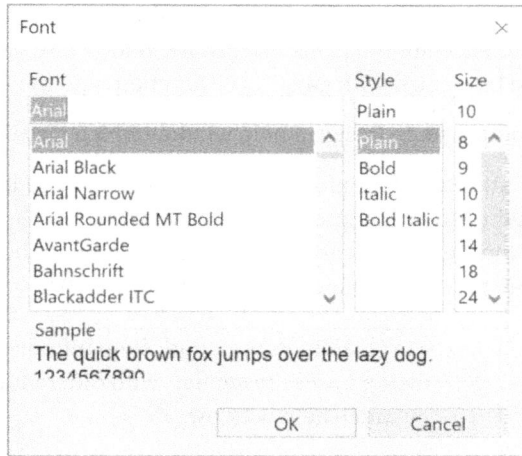

(3) **Font setting dialog box.** A standard font setting dialog box can be opened with the function uisetfont(), with the syntax

```
hFont=uisetfont(hTx,strTitle)
```

where hTx is the handle of the font, while strTitle is the string of the window title. A dialog box shown in Figure 10.7(b) can be opened, with the default hTx being the current font. The font can be selected from the dialog box, and the returned variable hFont is the handle of the font selected.

```
>> s = uisetfont
```

The following information is displayed:

```
    FontName: 'Times New Roman'
  FontWeight: 'normal'
   FontAngle: 'normal'
   FontUnits: 'points'
    FontSize: 10
```

If the font properties, such as the font size, are to be changed, function set() can be used. It can also be set directly with a command such as s.Fontsize=12.

(4) Simple help information dialog box. In fact, the simple help information dialog boxes are virtually the same as the standard warning boxes and error boxes. The difference are the icons used in the dialog box. Simple help information dialog box can be created with helpdlg() function, and the syntax is similar.

**Example 10.7.** Design a help window which may display information of several lines.

**Solutions.** If the information with several lines is displayed, a string matrix can be used to store the information. Alternatively, strings in cell arrays can also be used, with each line represented by a cell. Therefore the following statements can be used to display a help window, shown in Figure 10.8:

```
>> h=helpdlg({'Help information: Help information of warning and error' , ...
            'dialog boxes are similar, with different icons' , 'Compare!'} , ...
            'A simple help window')
```

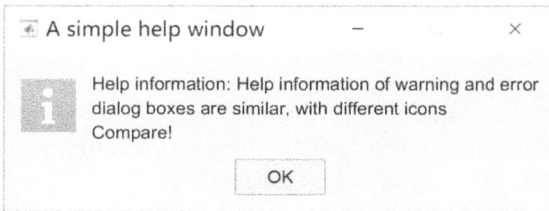

Figure 10.8: A simple help window.

(5) Online help window. Online help information of an existing function can be obtained in MATLAB with the function helpwin, having the syntax

helpwin function name

**Example 10.8.** Use help window to display information of a function.

**Solutions.** For instance, if one wants to find online help information of the function lyap(), the command helpwin lyap can be used to open a help window, and the help information of the function lyap() can be displayed directly in the help window.

```
>> helpwin lyap
```

## 10.2 Fundamental controls in interface design

Dialog boxes are the most widely used windows. If the user wants to interact with the computer, dialog boxes are the most important media. With the use of dialog boxes, commands can be sent to computers, parameters can be assigned to computers, and results can be fed back to the user. Fundamental controls and dialog boxes are illustrated in this section.

Controls in the interfaces can be constructed with function `uicontrol()`, and the syntax of the function is

`hctrl=uicontrol(property 1,value 1,property 2,value 2,...)`

A function call creates a control in the current window. With the function call, the properties of the control can be set. In the following presentation, the controls in dialog boxes are demonstrated.

### 10.2.1 Commonly used controls supported

The so-called dialog box is a temporally appearing window, displaying certain information or requesting the users to supply information. There are various controls in a dialog box, prompting the users to take certain actions. Dialog boxes are, in fact, the direct interface between the user and computer. Some of the commonly used controls in MATLAB are presented with necessary explanations. Each control has its own style, and various styles are illustrated as follows:

- Static text, with style name `'text'`, also known as a label. A static text control displays information in a dialog box. Usually it is used to display prompts or other necessary information for the users. If the dialog box requires the user to enter a variable, a static text control is usually assigned to prompt the user the meaning of the variable.
- Edit box, with style name `'edit'`. An edit box is a box with or without information in it. It allows the user to fill in or modify information inside it. The information in the edit box can be read by other controls.
- Frame, with style name `'frame'`. Frames are, in fact, a kind of decoration in the interfaces. They group sets of similar controls together. No genuine action is assigned in the frames.
- Listbox, with style name `'list'`. Many options are provided in a listbox. If there are too many options, a scroll bar appears automatically so that an option can be selected easily from it. In some special cases, in order not to occupy too much room, a listbox can be represented as a single row, with an arrow button appearing on its right. If the button is clicked, the listbox can be expanded. This kind of listbox is also known as a popup listbox.
- Scroll bar, with style name `'slider'`. A scroll bar allows the user to adjust the value of a variable in a visual way, by adjusting the positions of the indicator.
- Button, with style name `'pushbutton'`. Buttons are the most widely used controls in dialog boxes. Normally there is at least one button in a dialog box. Characters usually appear on the buttons to indicate their actions, for instance, OK, Yes, and Help buttons.

- Toggle button, with style name 'toggle'. Two states are allowed in a toggle button, pressed and released. A single click at the button toggles its status.
- Radio button, with style name 'radio'. A group of radio buttons can be used together. The user can select one and only one of the radio buttons. The selected one with a black dot in the button, and others are unselected. This relationship is referred to as mutually exclusive, which is not enabled in MATLAB. This relationship should be assigned by callback functions.
- Checkbox, with style name 'check'. The facilities of checkboxes are similar to those of the radio buttons, and the difference is that more than one option in a group can be selected at a time.
- Popup listbox, with style name 'popup', as explained in listbox.

**Example 10.9.** Create a listbox with options Item 1, my item 2 and test 3.

**Solutions.** A listbox object can be created with the following statements. One way is to describe the options with a string, with the options separated by vertical bars.

```
>> uicontrol('style','list','tag','mylist','Position',...
     [10,10,100,50],'String','Item 1|my item 2|test 3')
```

Another method is to use a cell array to describe the options, and the results are the same as those given above.

```
>> uicontrol('style','list','tag','mylist','Position',...
     [10,10,100,50],'String',{'Item 1','my item 2','test 3'})
```

### 10.2.2 Commonly used properties in controls

The classification and explanation of controls are presented in this section. The commonly used controls are illustrated, and some of the properties are the same in many controls:
- Units and Position properties. The definitions are the same as those of a window object. The reference point is the upper left corner of the window.
- String property. It is used to label the captions on the controls, or used as prompts.
- CallBack property. This is the most important property in graphical user interface design. Its property value is a string which can be executed directly. When the object is selected or changed, the string will be executed automatically. Normally, if an object is activated, a response function is executed automatically. The function is referred to as a callback function.
- Enable property, with the options 'on' and 'off', indicating whether the control is enabled or not.

- CData property, storing the true color bitmap image in a thee-dimensional array. The image appears on the controls such that the interface can be made colorful and more informative.
- TooltipString property, being a string to display prompts. When the mouse pointer is moved to a control, no matter whether the mouse button is pressed or not, the prompt is displayed.
- Interruptible property with options 'on' and 'off', indicating whether the callback function is interruptible or not.
- The properties related to fonts such as FontAngle and FontName.

### 10.2.3 Getting the handles

If one wants to let one control manipulate another control, the most important step is to find the handle of the control to be manipulated. Only the handle found can be manipulated. Normally, function findobj() can be used to find the handle of an object, with the following syntax:

h=findobj('property name', 'property value')

The function can be used to find all the handles matching the "property name" and "property value". In practical applications, in order to find accurately the desired handle, the Tag property of the object must be assigned into a unique string. If the handle is selected, functions set() and get() can be used to manipulate the properties in the control.

Some special functions are provided in MATLAB to extract the handles. For instance, command h=gcf can be used to extract the handle of the current figure window, and return the handle to the variable h. Function gca is used to extract the handle of the current axis, while function gco can be used to extract the handle of the current object. Function gcob can be used to extract the handle of the object whose callback function is activated.

**Example 10.10.** Write a MATLAB function such that if it is executed, a figure window can be opened. In the window there are two controls. One of them is a button, the other is a static text. When the button is clicked, the string "Hello World!" is displayed in the static text control.

**Solutions.** This is a simple application in object oriented programming. As illustrated before, function figure() opens a blank window. Function uicontrol() can be used to draw the two controls in the window. How can we implement the expected responses? From the point of view of object oriented programming, a callback function can be designed for the button. Its actions are assigned as: (1) find the handle of the static text control, (2) set the String property to "Hello World!".

It is a crucial step in finding the handle of the static text. Function `findobj()` can be used to find the handle of the object. Then function `set()` can be used to assign its **String** property. In order to efficiently use the function `findobj()`, a unique tag must be assigned to the static text object. Therefore, the following statements can be used to design the interface:

```
>> h=figure; pos=[0,100,100,20];
   uicontrol('style','text','Tag','txtHello','Position',pos)
   uicontrol('style','pushbutton','String','OK',...
            'Callback',['h0=findobj(''Tag'',''txtHello'');',...
                        'set(h0,''String'',''Hello World!'')'])
```

We were trying to avoid talking about detailed setting of the properties of windows and controls, since setting their positions and sizes cannot be done in a straightforward manner. The default values are used all the time. The position of the static text control is adjusted by the statement. Otherwise the two controls may overlap. If a visual design tool is used, the design may become much simpler and more straightforward.

## 10.3 Graphical user interface design tool – Guide

Command `guide` can be issued in the MATLAB command window. The interface shown in Figure 10.9 is opened, to prompt the user to select an appropriate interface from the listed GUI templates, where an empty interface can be opened with the **Blank GUI (Default)** option. The other interfaces are **GUI with Uicontrols**, **GUI with Axes and Menu**, and **Modal Question Dialog**. Also one is allowed to choose **Open Existing GUI**.

Here a blank window is needed in the interface design. Selecting the **Blank GUI** template, then clicking the **OK** button, the program design interface in Figure 10.10

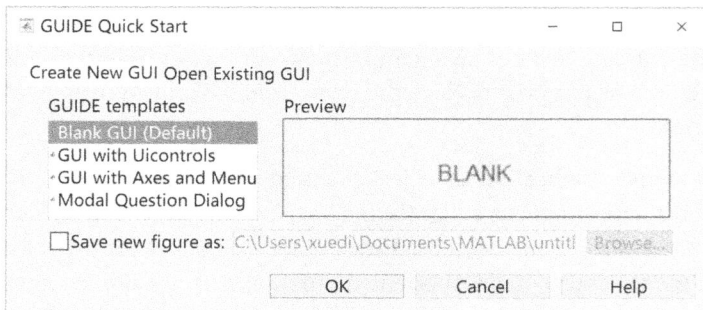

**Figure 10.9:** The main interface of Guide program.

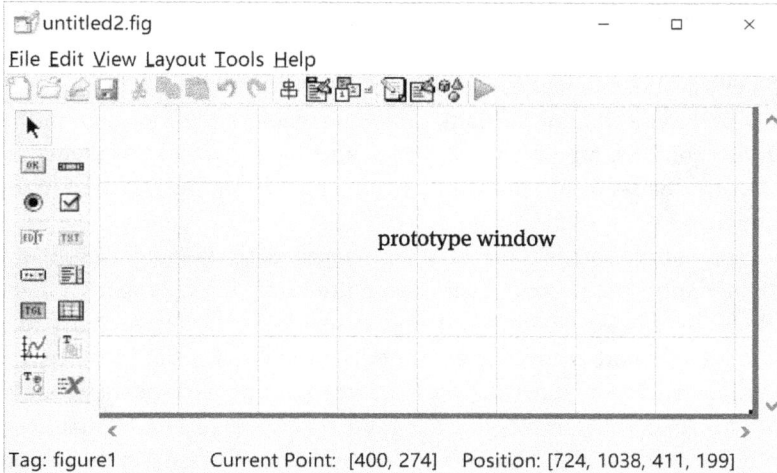

**Figure 10.10:** GUI editing interface.

is opened, where the right-hand side region is the prototype of the window to be designed.

In the left-hand side of the interface, various controls are provided as shown in Figure 10.11. Selecting a control from the list, it can be drawn in the prototype window. Various controls can be drawn in the prototype window. The expected interface can be designed in this way. The fundamental knowledge of handle graphics is introduced, and user graphical interface programming is demonstrated with simple examples.

**Example 10.11.** Design an interface with Guide for Example 10.10.

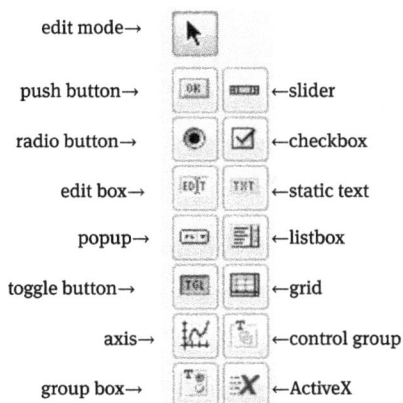

**Figure 10.11:** Guide controls.

**Solutions.** A blank window can be drawn with Guide. Draw a static text control and a push button. When the button is pressed, the text control appears with the string "Hello World!". The design procedures are as follows.

(1) Draw the prototype window. A blank prototype window can be brawn in the interface, with two controls, a push button and a label (or static text), as shown in Figure 10.12(a). It can be seen that the window and controls can be adjusted in an interactive manner.

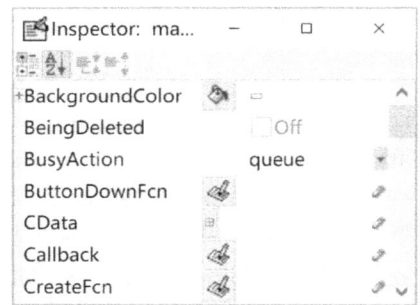

(2) Modify the properties of the controls. The properties of the label control should be modified to an empty string. Double click the control to open its property dialog box. Modify the **String** property to an empty one. Also, assign the **Tag** property to a meaningful one, such that other controls may find its handle easily. For instance, set it to txtHello, as shown in Figure 10.12(b). Note that, when setting the tags, they should be set to unique strings in the interface, such that other controls can find it uniquely, rather than finding others. For the convenience of programming, the tag of the button can be set to btnOK.

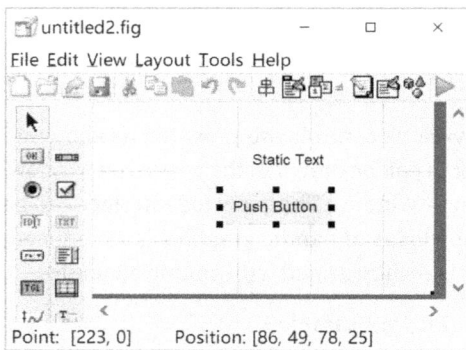

(a) draw the necessary controls   (b) modify the properties

**Figure 10.12:** Design and modification of the interface.

(3) Generate automatically the framework. With the prototype window drawn by Guide, the interface can be saved to a *.fig file, such as c10exgui1.fig. Meanwhile, another file c10exgui1.m can be automatically generated, with the main framework shown below:

```
function varargout = c10exgui1(varargin)
gui_Singleton = 1;
gui_State = struct('gui_Name',        mfilename, ...
                   'gui_Singleton',  gui_Singleton, ...
                   'gui_OpeningFcn', @c10exgui1_OpeningFcn, ...
                   'gui_OutputFcn',  @c10exgui1_OutputFcn, ...
```

```
                   'gui_LayoutFcn',  []  , ...
                   'gui_Callback',   []);
if nargin && ischar(varargin{1})
    gui_State.gui_Callback = str2func(varargin{1});
end
if nargout
    [varargout{1:nargout}] = gui_mainfcn(gui_State, varargin{:});
else
    gui_mainfcn(gui_State, varargin{:});
end
% End initialization code - DO NOT EDIT
function c10exgui1_OpeningFcn(hObject,eventdata,handles,varargin)
handles.output = hObject;
guidata(hObject, handles);
function varargout=c10exgui1_OutputFcn(hObject,eventdata,handles)
varargout{1} = handles.output;
function btnOK_Callback(hObject,eventdata,handles)  % the blank framework
```

(4) **Write the callback functions.** Analyzing the requests, it can be seen that the needed response function of the button is, when the button is pressed, setting the **String** property of the label control to "Hello World!". A callback function for the button should be designed. Since the tag of the label control is **txtHello**, its handle is then **handles.txtHello**. The following callback function can then be written as:

```
function varargout = btnOK_Callback(hObject, eventdata, handles)
set(handles.txtHello,'String','Hello World!');
```

Compared with the code in Example 10.10, although the code is a little complicated, some troubles of Example 10.10 are avoided, for instance, the positions of the window and controls, as well as the adjustments of the sizes and their units. This information can be set with the Guide program in an interactive way. Besides, the structure of the callback function is more formal, and can be used in more complicated programming.

**Example 10.12.** MATLAB graphical user interface design is, in fact, an object-oriented programming method. If in a figure window a three-dimensional surface is expected, the target in the prototype in Figure 10.13 is given. The fundamental requirements for the interface are:
(1) Drawing the main axis, for the final three-dimensional surface;
(2) An edit box, ready to accept data or function for the plot;
(3) Two buttons, one for plot drawing, the other for demonstration;
(4) A group of three edit boxes, for setting the position of the light source;

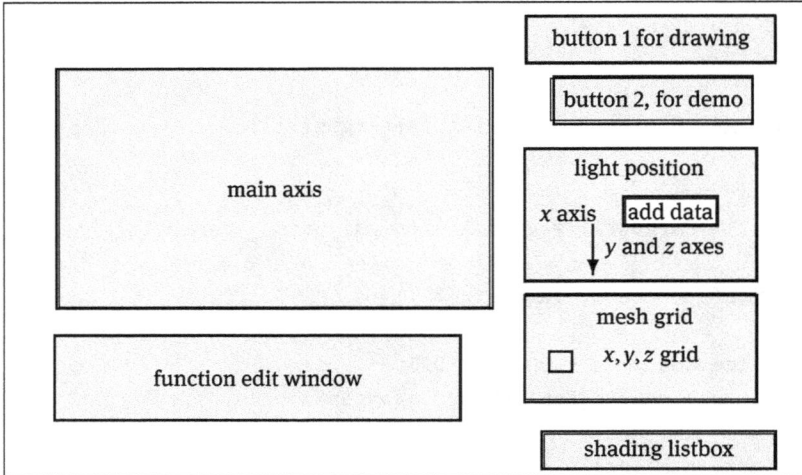

**Figure 10.13:** A prototype of the interface.

(5) A set of three checkboxes, to determine whether grids are needed;
(6) A listbox to change the options of shading command.

**Solutions.** Based on the above assumptions, the prototype of the interface can be drawn directly with Guide program, as shown in Figure 10.14. The edit boxes and checkboxes can further be aligned by the tools ⌖.

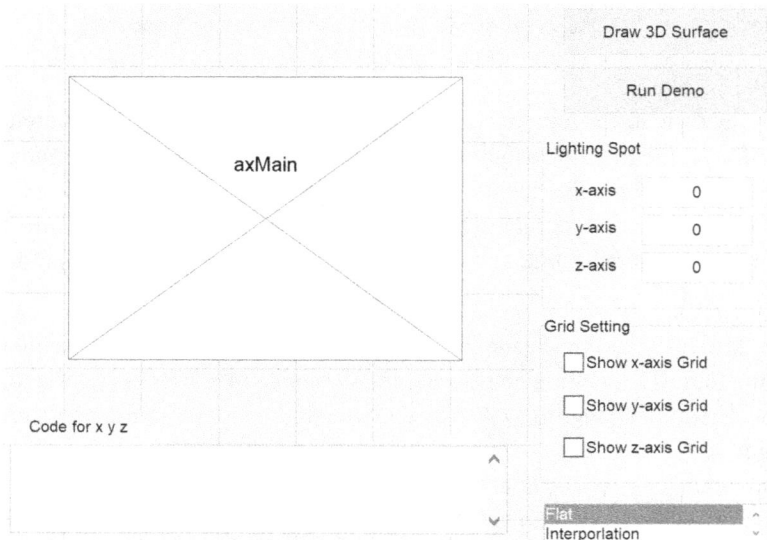

**Figure 10.14:** Prototype windows designed in Guide.

The tasks can be assigned for several different controls, as shown in Figure 10.15. It can be seen that zones A and B are not assigned any tasks. They can be used to accept data and draw plot. Their handles are very important. For the easy acquirement of their handles, their Tag properties should be set to meaningful ones, as axMain and edtCode, respectively. Meanwhile, the edtCode should be set to enable multiple lines of strings, by setting its Max property to a value larger than 1. For instance, it can be set to 100.

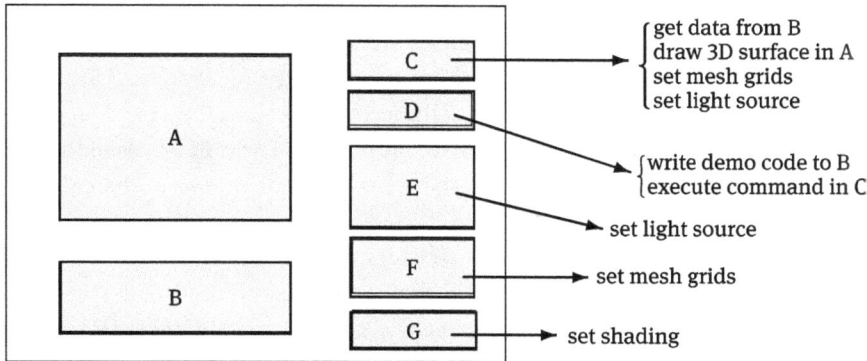

**Figure 10.15:** Illustration of the assigned tasks.

The tags of the controls can be set as follows:
- The tags of buttons in zones C and D are set to btnDraw, btnDemo.
- The tags of the 3 edit boxes in zone E can be set to edtX, edtY, and edtZ.
- The tags of the 3 check boxes in zone F can be set to chkX, chkY, and chkZ.
- The tag in the listbox in zone G is set to lstFill, with its String property set to Flat ↵ Interpolation ↵ Faceted, by clicking the 📰 button beside the lstFill object, where ↵ represents carriage return key.

With the tasks assigned, the main interface can be designed, in the function c10fgui3(), and the listing of the automatically generated framework is given below:

```
function varargout = c10eggui3(varargin)
gui_Singleton = 1;
gui_State = struct('gui_Name',        mfilename, ...
                   'gui_Singleton',  gui_Singleton, ...
                   'gui_OpeningFcn', @c10eggui3_OpeningFcn, ...
                   'gui_OutputFcn',  @c10eggui3_OutputFcn, ...
                   'gui_LayoutFcn',  [] , ...
                   'gui_Callback',   []);
```

```
if nargin && ischar(varargin{1})
    gui_State.gui_Callback = str2func(varargin{1});
end
if nargout
    [varargout{1:nargout}]=gui_mainfcn(gui_State, varargin{:});
else
    gui_mainfcn(gui_State, varargin{:});
end
```

It can be seen that the main program generated is exactly the same as that generated in other interfaces. For the two different programs, what are the differences in the interface description? Details of interface description are stored in the .fig file. There is no difference at all in the main framework of the *.m file. The differences in the *.m files lie in the responses to the controls, where the implementation of the callback functions are different in the interface.

For the tasks assigned to zone C, a callback function can be written for the button btnOK. The assigned tasks are: extract the string from the edit box edtCode; evaluate the string to get data; and draw the three-dimensional surface in the axis axMain. The callback function can be written as follows:

```
function btnDraw_Callback(hObject, eventdata, handles)
try
    str=get(handles.edtCode,'String'); str0=[];
    for i=1:size(str,1), str0=[str0, deblank(str(i,:))]; end
    eval(str0); axes(handles.axMain); surf(x,y,z);
catch, errordlg('Error in code'); end
```

Note that the trial structure try ... catch is written to prevent errors in the function edit box. If there are unrecognized or inexecutable errors in the string, an error dialog box is opened.

Let us consider the callback function of button btnDemo in zone D. The tasks assigned to the button are: set the demonstration MATLAB code to edit box edtCode, then call the callback function of button btnDraw directly to display the three-dimensional surface. The callback function for the button is

```
function btnDemo_Callback(hObject, eventdata, handles)
str1='[x,y]=meshgrid(-3:0.1:3, -2:0.1:2);';
str2='z=(x.^2-2*x).*exp(-x.^2-y.^2-x.*y);';
set(handles.edtCode,'String',str2mat(str1,str2));
btnDraw_Callback(hObject, eventdata, handles)
```

Let us think about the callback function of zone E. There are three edit boxes in zone E, used for setting the position of the light source. Three callback functions are considered for the controls edtX, edtY, and edtZ. The values can be extracted from the three edit boxes, and then the position of the light source can be set in the axis axMain. Therefore, the following callback function can be written:

```
function edtX_Callback(hObject, eventdata, handles)
try
    xx=str2num(get(handles.edtX,'String'));
    yy=str2num(get(handles.edtY,'String'));
    zz=str2num(get(handles.edtZ,'String'));
    axes(handles.axMain); light('Position',[xx,yy,zz]);
catch, errordlg('Wrong data in Lighting Spot Positions'); end
```

For the convenience in programming, there is no need to write the callback functions. The function edtX can be used as a common function, i. e., let edtY object call the common function directly as follows. Also, the callback function of object edtZ can be written in the same manner.

```
function edtY_Callback(hObject, eventdata, handles)
edtX_Callback(hObject, eventdata, handles)
```

Similarly, the callback function of zone F can be written as

```
function chkX_Callback(hObject, eventdata, handles)
xx=get(handles.chkX,'Value'); yy=get(handles.chkY,'Value');
zz=get(handles.chkZ,'Value');
set(handles.axMain,'XGrid',onoff(xx),'YGrid',onoff(yy),...
    'ZGrid',onoff(zz))
% ---    the user defined subfunction
function out=onoff(in)
out='off'; if in==1, out='on'; end
```

The status of the chkX is measured. Based on the result, the status of *x* axis mesh grids can be set. Since there is no possibility of errors in the check boxes, there is no need to implement it with the try ... catch structure. A subfunction onoff() can be written to convert 0 or 1 into strings 'off' or 'on'. Another callback function chkY_Callback() should also be written, letting it call directly the same chkX_Callback function.

For the controls in the listbox in zone G, the callback function extracts the option from the listbox, and the appropriate shading command can be issued to get the final shading results. The callback function can be written as follows:

```
function lstFill_Callback(hObject, eventdata, handles)
v=get(handles.lstFill,'Value'); axes(handles.axMain);
switch v
    case 1, shading flat;    % surface with no mesh grids
    case 2, shading interp;  % interpolations with smooth surface
    case 3, shading faceted; % surface with mesh grids
end
```

With the program in file `c10eggui3.m`, the interface shown in Figure 10.16 can be opened.

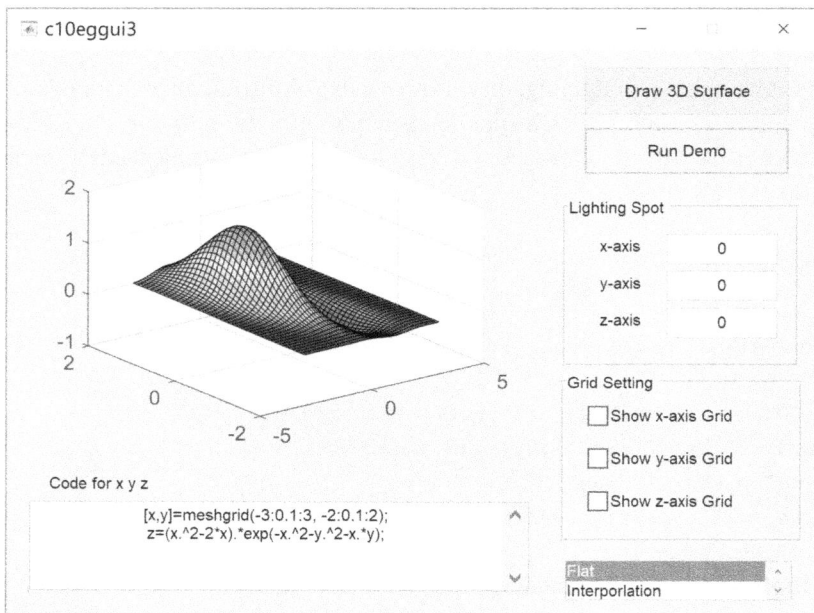

**Figure 10.16:** The effect when Run Demo button is clicked.

With the interfaces presented earlier, it can be seen that each interface is processed individually. If an interface has several subinterfaces, variables are usually transferred among different interfaces. A good selection is that the data can be transferred through the **UserData** property of the main interface. The implementation of variable transfer will be demonstrated through the following example.

**Example 10.13.** Write a simple matrix processing interface. A dialog box can be opened to input special matrices, and the results can be displayed in the interface.

**Solutions.** An interface can be designed in the form shown in Figure 10.17, with the tags of some of the controls assigned as in the figure. The tasks assigned to the active controls are summarized below:

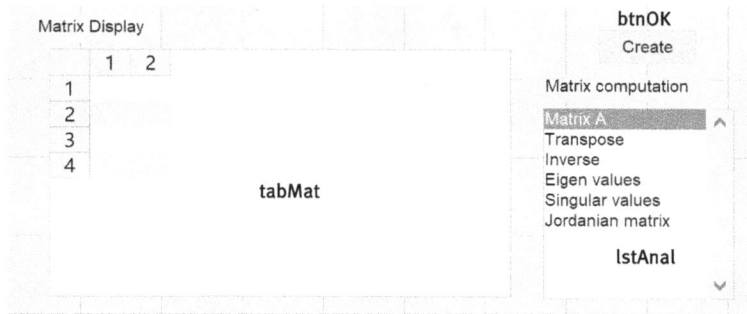

**Figure 10.17:** Design of a matrix analysis interface.

(1) **Create button.** If the button is pressed, a low-level dialog box can be displayed. More specifically, the callback function examines whether such a dialog box is opened or not. If it is opened, set it to visible mode. The MATLAB implementation of the callback function is

```
function btnOK_Callback(hObject, eventdata, handles)
h=findobj('Tag','figA');
if length(h)==0, c10exmat1; else, set(h,'Visible','on'); end
display_mat(handles)
```

(2) **The listbox on the right.** It can be used to open the matrix to be displayed, with MATLAB implementation being

```
function lstAnal_Callback(hObject, eventdata, handles)
display_mat(handles)
```

Since the two controls are executing the task of displaying a matrix, a separate subfunction can be written, whose assigned task is reading matrix $A$ from the UserData property of the interface; the type of the matrix is to be displayed from the listbox; with the instructions in the listbox, computing the matrix $A_1$ to be displayed; writing a separate subfunction $A_1$ and an interface shown in Figure 10.18.

```
function display_mat(handles)
A=get(gcf,'UserData'); key=get(handles.lstAnal,'Value');
switch key,
```

```
   case 1, A1=A; case 2, A1=A.'; case 3, A1=inv(A);
   case 4, A1=eig(A); case 5, A1=svd(A); case 6, A1=jordan(A);
end
set(handles.tabMat,'Data',A1)
```

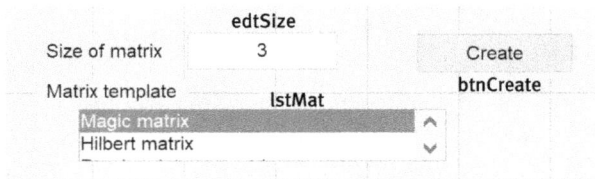

**Figure 10.18:** The subinterface designed for special matrices.

For the subinterface, the only active control is the Create button. The tasks are assigned as follows: extract the order $n$ from the edit box; select the type of the matrix from the listbox; generate the appropriate matrix $A$; find the handle of the main interface, and assign $A$ to the UserData property of the main interface.

```
function btnCreate_Callback(hObject, eventdata, handles)
n=eval(get(handles.edtSize,'String'));
key=get(handles.lstMat,'Value');
set(gcf,'Visible','off'); h=findobj('Tag','figMain');
switch key
   case 1, A=magic(n); case 2, A=hilb(n);
   case 3, A=randi([-1,1],n); case 4, A=pascal(n);
   case 5, A=gallery('binomial',n); case 6, A=gallery('gcdmat',n);
end, set(h,'UserData',A);
```

For such a problem, the best way to "close" a dialog box is to set it to an invisible one, rather than really closing it.

## 10.4 Advanced techniques in interface design

The powerful graphical interface design tool in MATLAB, Guide, can be used to design various GUIs. Apart from the dialog boxes discussed so far, commonly used interface tools such as menu systems, toolbars, and shortcut keys can also be designed. Besides, powerful ActiveX controls can also be embedded in the interface. In this section, these topics are covered in this section.

### 10.4.1 Design of menu systems

With the powerful facilities provided in Guide program, ordinary dialog boxes can be designed easy; besides, a complicated menu system can be designed, too. The menu system can also be designed with the Guide program, with a design interface shown in Figure 10.19(a). The controls such as **Object Align, Grid and Ruler** can be used to design menu system. Toolbars can also be designed by the interface.

Selecting **Tools → Menu editor** command, the menu editor in Figure 10.19(b) can be can be opened. The menu shown in Figure 10.20 can easily be created, and the method is straightforward. Shortcut keys can also be assigned with the menu editor. The whole program can be saved as `c10eggui2.m` file.

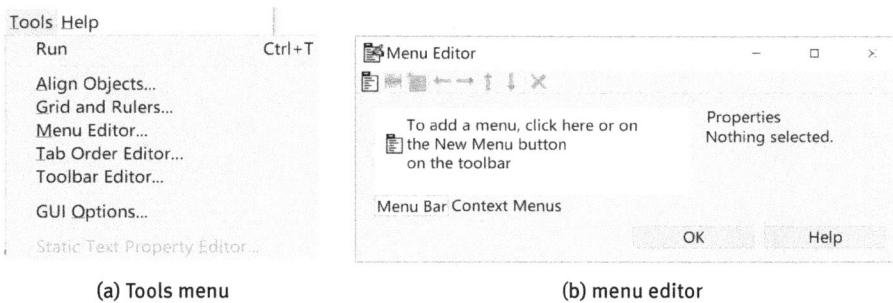

(a) Tools menu           (b) menu editor

**Figure 10.19:** Tools menu and menu editor.

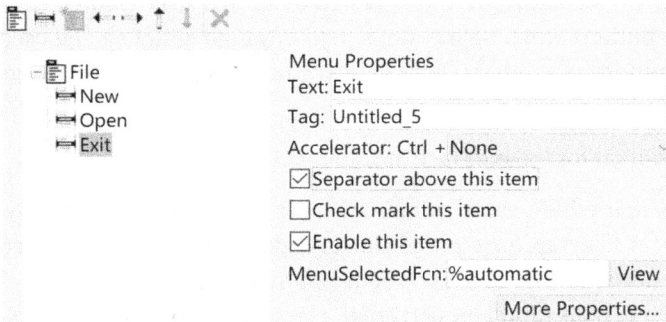

**Figure 10.20:** Menu editor.

### 10.4.2 Design of toolbars

Toolbars can be designed in MATLAB interfaces. The Tools → Toolbar Editor menu item in Guide can be selected to open the toolbar editor, as shown in Figure 10.21.

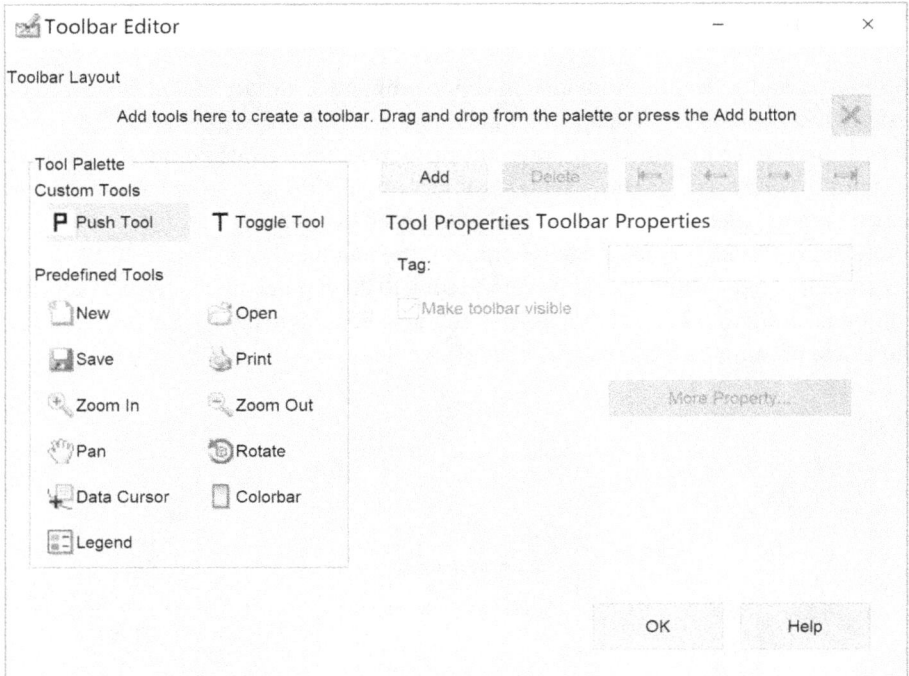

**Figure 10.21:** Toolbar editor interface.

The icons in the toolbar can be assigned to predefined ones, and user-defined ones are also supported. The buttons and tangle buttons in the Toolbar editor can be used to create user-defined buttons in toolbars, with their CData property set to an image variable.

**Example 10.14.** Add a toolbar for the interface in Example 10.12.

**Solutions.** Open the toolbar editor, some commonly used tools can be selected, by double clicking the icon, or selecting an icon and clicking Add button. If a user-defined button in the toolbar is needed, then click Push Tool and then Add button. An image variable *W* can be assigned to the CData property of the button. For instance, the toolbar in Figure 10.22 can be developed.

When the program is edited, the standard icons inherit the facilities in their parent icons. For instance, the points on the curves can be read with the cursor icon, and there is no need too write a program for it. For the two user-defined buttons, the following callback functions can be written:

```
function tolXZoom_ClickedCallback(hObject, eventdata, handles)
zoom xon
```

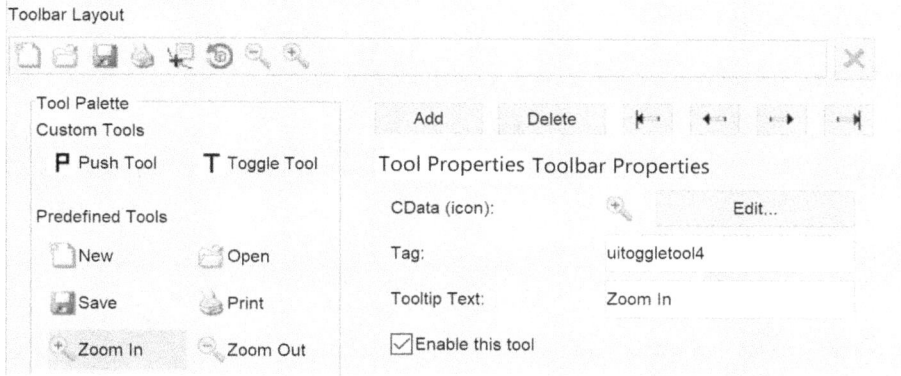

**Figure 10.22:** User-defined toolbar design.

```
function tolYZoom_ClickedCallback(hObject, eventdata, handles)
zoom yon
```

Even simpler, there is no need to write callback function in this case. Clicking the ⊘ button, opening its property editor interface, and filling 'zoom xon' command directly in the ClickedCallback edit box, the problem can be solved.

### 10.4.3 Embedding ActiveX controls

ActiveX usually refers to reusable components, such as the Windows Media Player or database components, developed by software providers such as Microsoft. The ActiveX components can be adopted and embedded in the MATLAB GUIs to make the interfaces more powerful, without the need for low-level programming. These ActiveX components can be embedded in MATLAB GUIs by clicking the ⊠X button, then a dialog box shown in Figure 10.23 will be displayed, allowing the user to select a proper ActiveX component from the listbox. An example is given below to show how to use ActiveX in GUI design.

**Example 10.15.** Design a multimedia file player interface with MATLAB.

**Solutions.** It seems that the implementation of the task is rather difficult. It is indeed difficult to implement the facilities with low-level statements. Luckily, the ActiveX controls can be used directly in MATLAB. If an appropriate ActiveX control for a multimedia file player can be found, the problem can be solved immediately.

Command guide can be used to open a blank GUI window. Draw an ActiveX control in the window. A dialog box appears with the list of available ActiveX controls, from which the Windows Media Player component can be selected. A button Create

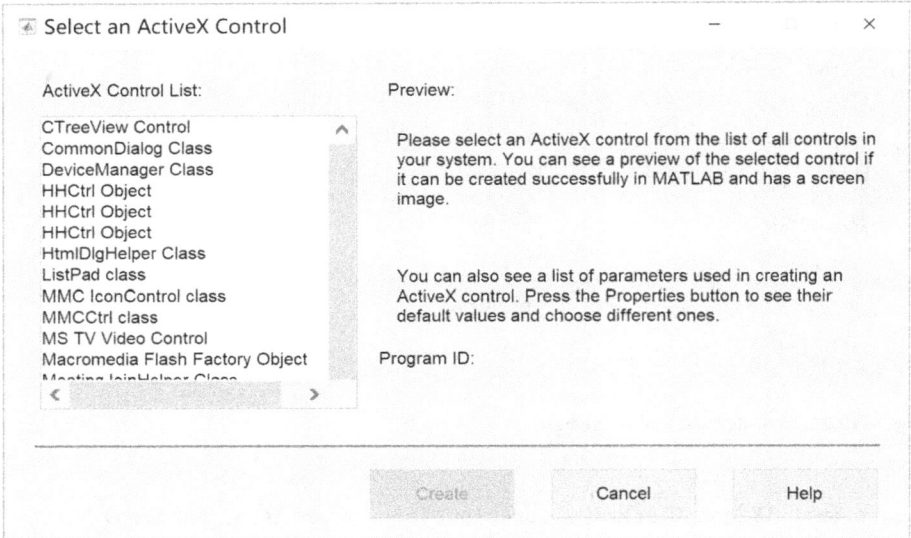

**Figure 10.23:** ActiveX control selections.

can be clicked, and the standard Windows Media Player control appears. The default Tag of the control is **activex1**. A button can be added beside the control, and the **Tag** can be set to **btnLoad**, with the **String** property cleared.

The image of the icon ⬚ can be obtained, and saved into file **btnFile.bmp**. It can be loaded into MATLAB workspace with **W**=imread('btnFile.bmp'). The variable **W** can be assigned to the **CData** property of the button. The prototype window in Figure 10.24 can be created. The following callback function for the button can be written:

```
function btnLoad_Callback(hObject, eventdata, handles)
[f,p]=uigetfile('*.*','Select a media file');
if f~=0, set(handles.activex1,'URL',[p,f]); end
```

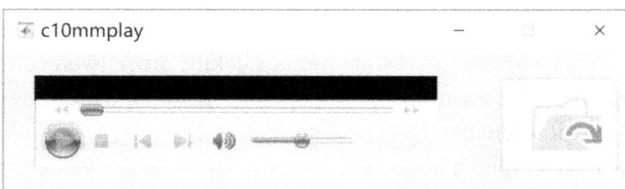

**Figure 10.24:** Multimedia file player interface.

If the button is clicked, a multimedia file can be loaded through the standard file name input dialog box, and the file can then be played directly.

It can be seen from the example that, with the powerful facilities and potentials, various programs can be written with simple MATLAB commands.

## 10.5 APP packaging and publication

If the user wants to publish a MATLAB toolbox or a set of MATLAB functions he/she designed, the code should be packaged first. The correct procedures are as follows: pressing the APP→APP Packaging button, a dialog box shown in Figure 10.25 appears. Filling in all the necessary information, and associating all the files he/she wants to package, by clicking Package button, a packaged file with extension of .mlappinstall is created. The file is ready to be installed automatically in MATLAB environment.

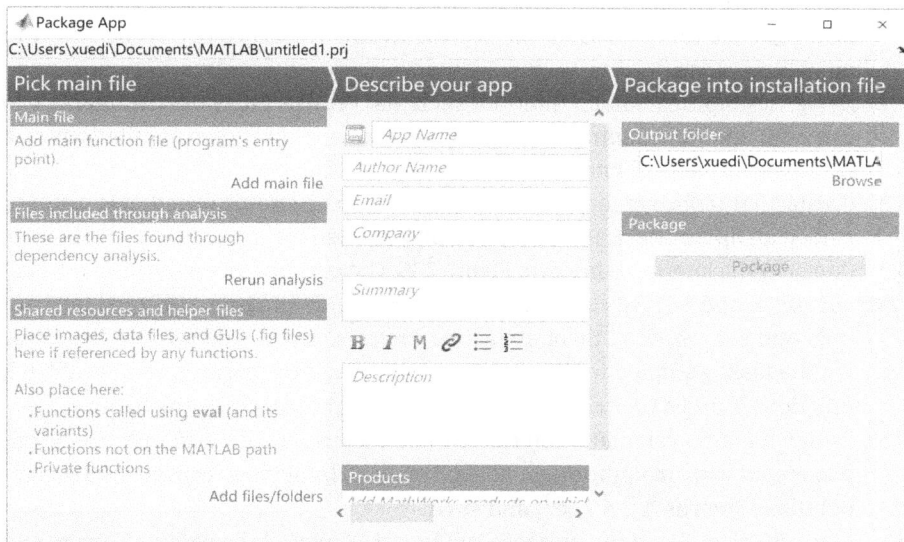

**Figure 10.25:** APP packaging interface.

## 10.6 Problems

10.1   Design a simple user interface, with an edit box. If the radius of a circle is entered in the edit box, then pressing the return key, the area and volume of the circle or a ball should be selected automatically.

10.2 Considering again the interface in Problem 10.1, fault tolerance facilities can be added. When the string entered in the edit box is not a number, an error message should be displayed. Otherwise, the area and volume of the circle or ball can be computed.

10.3 Design a temperature converting interface, where two edit boxes are provided side-by-side. The boxes are ready to accept temperatures in Celsius and Fahrenheit, respectively. If one of them is changed, the other box gives the converted temperature. Hints: the conversion formula is $C = 5(F - 32)/9$.

10.4 Design an interface which allows the user to assign magnitude, frequency, and initial phase, and by a click of a button, the sinusoidal curve should be drawn automatically in the window. A listbox can also be designed to allow the user select from the list what kind of curve to plot. In the listbox, the sine, cosine, and tangent curves can be selected.

10.5 For the interface designed in Problem 10.4, add a scroll bar which allows the user to adjust continuously the magnitude. Also the scroll bar and edit box should be associated, so as to implement the same actions.

10.6 Design a simple interface where the users may specify the initial speed and vertical distance to the ground. Show dynamically the trajectory of the object in animation mode.

10.7 Design a simple calculator interface such that the users may use the interface to carry out addition, subtraction, multiplication, and division computations as if using a hardware calculator.

10.8 Create an interface to display clock time, like an electronic watch. It is better to embed an appropriate ActiveX control in the interface.

10.9 Consider the interface shown in Figure 7.35. The user may open the files `vol_visual4d.m` and `vol_visual4d.fig`, with the Guide program. Observe the controls and see which of them are active controls, and which are passive. What are the tasks assigned to the active controls? Read the corresponding callback functions. Build a toolbox for the interface and then package the toolbox.

10.10 Design a matrix calculator interface where the display of the matrix can be implemented with an appropriate ActiveX control. Implement certain matrix computations such as $A^n$, $A^{-1}$, $e^A$, and $\sin A$.

10.11 In practical applications, suppose you found a curve while reading a paper, and you only have the electronic version of the paper. Suppose you want to restore the curve in vector graphs, rather than in bitmaps. You need to extract the key points in the curve. Create a user interface and display initially the bitmap of the image. Besides, users should be allowed to select the axis from the image, and to locate the points of interest from the curve such that the data in the curves can be extracted.

# Bibliography

[1] Anderson E, Bai Z, Bischof C, et al. LAPACK Users' Guide. Philadelphia: SIAM Press, 1999.

[2] Atherton D P, Xue D. The analysis of feedback systems with piecewise linear nonlinearities when subjected to Gaussian inputs. In Kozin F, Ono T, eds. Control Systems, Topics on Theory and Application, 23–38. Tokyo: Mita Press, 1991.

[3] Bodin P. PLOTY4 support for four y axes. MATLAB Central File ID: #4425, 2004.

[4] Dongarra J J, Bunsh J R, Molor C B. LINPACK User's Guide. Philadelphia: Society of Industrial and Applied Mathematics, 1979.

[5] Garbow B S, Boyle J M, Dongarra J J, et al. Matrix Eigensystem Routines – EISPACK Guide Extension. Lecture Notes in Computer Sciences, vol. 51. New York: Springer-Verlag, 1977.

[6] Gilbert D. PLOTXX create graphs with two x axes. MATLAB Central File ID: #317, 1999.

[7] Gilbert D. Extended plotyy to three y-axes. MATLAB Central File ID: #1017, 2001.

[8] Gomez C, Bunks C, Chancelier J-P, et al. Engineering and Scientific Computing with Scilab. New York: Springer, 1999.

[9] Gongzalez R C, Woods R E. Digital Image Processing. Englewood Cliffs: Prentice-Hall, 2nd edition, 2002.

[10] Lamport L. LaTeX: A Document Preparation System – User's Guide and Reference Manual. Reading MA: Addision–Wesley Publishing Company, 2nd edition, 1994.

[11] Majewski M. MuPAD Pro Computing Essentials. Berlin: Springer, 2nd edition, 2004.

[12] Moler C. Experiment with MATLAB. Beijing: BUAA Press, 2013.

[13] Monagan M B, Geddes K O, Heal K M, et al. Maple 11 Advanced Programming Guide. Waterloo: Maplesoft, 2nd edition, 2007.

[14] Numerical Algorithm Group. NAG FORTRAN Library Manual, 1982.

[15] Register A H. A Guide to MATLAB Object-Oriented Programming. Boca Raton: Chapman & Hall/CRC, 2007.

[16] Smith B T, Boyle J M, Dongarra J J, et al. Matrix Eigensystem Routines – EISPACK Guide. Lecture Notes in Computer Sciences. New York: Springer-Verlag, 2nd edition, 1976.

[17] Wolfram S. The Mathematica Book. Champaign: Wolfram Media, 5th edition, 2003.

[18] Xue D Y. Mathematics education made more practical with MATLAB. Presentation at the First MathWorks Asian Research Faculty Summit, Tokyo, Japan, November, 2014.

[19] Xue D Y. Fractional-order Control Systems – Fundamentals and Numerical Implementations. Berlin: de Gruyter, 2017.

[20] Xue D Y. FOTF Toolbox, MATLAB Central File ID: #60874, 2017.

[21] Xue D Y, Chen Y Q. Modeling, Analysis and Design of Control Systems Using MATLAB and Simulink. Singapore: World Scientific, 2014.

[22] Xue D Y, Chen Y Q. Scientific Computing with MATLAB. Boca Raton: CRC Press, 2nd edition, 2015.

https://doi.org/10.1515/9783110666953-011

# MATLAB function index

Bold page numbers indicate where to find the syntax explanation of the function. The function or model name marked by * are the ones developed by the authors. The items marked with ‡ are those down-loadable freely from Internet.

# Index